Oscilloscopes

Stan Prentiss

Reston Publishing Company, Inc.
A Prentice-Hall Company
Reston, Virginia

Library of Congress Cataloging in Publication Data

Prentiss, Stanton Rust.
  Oscilloscopes.

  Includes index.
  1. Cathode ray oscilloscope.   I. Title.
TK7878.7.P73   621.3815'48   80-21955
ISBN 0-8359-5354-8

© 1981 by
Reston Publishing Company, Inc.
A Prentice-Hall Company
Reston, Virginia 22090

All rights reserved. No part of this book
may be reproduced in any way, or by any
means, without permission in writing
from the publisher.

10 9 8 7 6 5 4 3 2 1

Printed in the United States of America

Table of Contents

Acknowledgments, ix

Preface, 1

1 **Oscilloscopes and Their Uses, 3**

Scope Basics, 3
Vertical Amplifiers, 6
Time Base, 9
Probes, 10
Calibration and AC-DC Measurements, 13
Oscilloscope Applications—Live, 15
Vectorscopes, 21
   In Color TV, 23; Troubleshooting, 26
Professional Oscilloscopes, 28

2 **Spectrum Analyzers, 34**

Applications of Spectrum Analyzers, 36
   Amplitude Measurements, 36; Resolution, 36; Frequency and Time Bases, 38
AM Spectrum Measurements, 40
   AM Modulation, 40; AM Distortion, 41; Harmonics and Spurs, 42; Signal/Noise Measurement at RF, 43; Oscillator Distortions, 43
FM Spectrum Measurements, 44
   FM Modulation, 44
Omni-Checks-Everywhere, 47
   Power Output Test, 47; Harmonic Distortion, 47; Intermodulation Distortion, 47; Signal-to-Noise (S/N) Ratio, 48; S/N Ratio with Dual Trace Scope and Signal Source, 49
Summary, 52

v

vi / Table of Contents

3 **Logic Analyzers, 53**

Analyzers Defined, 54
 Timing Analyzers, 55; State Analyzers, 55
Testing Procedures, 56
 Selective Triggering, 56; H-P's Digital Analysis, 58
How They Operate, 60
 Synchronous Sampling, 60; Asynchronous Sampling, 60; Triggering, 61; Word Recognition, 61; Cathode Ray Tube Displays, 62; Binary Display, 62; Map Display, 63
General Troubleshooting, 65
 A 7D01 Analyzer Plug-In, 65

4 **Storage/Sampling Oscilloscopes and Time Domain Relectometry, 70**

Storage Oscilloscopes, 71
Reasons for Storage, 73
Storage Features, 73
 Bistable Storage, 73; Bright Bistable, 73; Variable Persistence, 74; Fast Mesh Transfer, 75; Fast Multimode, 75
What Storage Scopes Do and Why, 76
Fast Scopes and Storage, 80
Sampling Oscilloscopes, 82
Time Domain Reflectometry, 83
Conclusions, 88

5 **Investigating Video Terminals and Cassettes, Including the Art of Using Vectorscopes, 89**

Chroma and the Vectorscope, 90
The Vector Pattern, 94
Gaited Rainbow Color Bar Generators, 96
Conclusions, 99
Video Cassettes, 100
Theory of Operation, 101
 Drawbacks, 104; Clock Timing, 106; Troubles and Pertinent Waveforms, 107

6 **Waveform Analysis, 112**

Basic Configurations, 112
 Square (or Rectangular) Waves, 113; Triangular Waves, 114; Sawtooth Waveshapes, 115; Staircases, 116

VITS, 117
VIRS, 120
Actual Waveforms, 122
  Fundamental Waveshapes, 125; Rectangular Voltages, 125; Sinusoidal Voltages, 128; Sawtooth or Ramp Waveforms, 131; Composite Video Waveforms, 132
Examples of Defective Waveforms and Why, 138
  Power Supply Ripple, 138; A Multi vibrator, 140; Time Constants, 141; Video Alignments, 143; Sync Problems, 145; AFC Phase Control, 149
Scope Calibration, 150
  Vertical Calibration, 150; Horizontal Calibration, 151

Index, 155

Acknowledgments

The author would like to acknowledge the great assistance of Tektronix, Hewlett-Packard, Sencore, and Dynascan Corp., for their own equipment-use tutorials as well as suggestions offered by specialists in these outstanding organizations. Whenever their material has been directly applied, due credit appears in this publication.

Preface

You, the reader or student, will find this a very different book from others on the market dealing with vital test equipment subjects. For only briefly are you told what an oscilloscope is, while you are shown in great detail what it does and why. You will see that all manner of applications are included, ranging from basic service procedures to highly complex design and evaluation studies.

Such subjects as digital logic, cable television, two-way radio, frequency and amplitude modulation, video terminals, high-speed waveshaping, time domain reflectometry, spectrum analysis, and storage techniques are all covered, often illustrating precise problems and how these are recognized and measured. There are special chapters devoted to VECTOR applications, using both sidelock and NTSC generators, as well as an intensive study of all fundamental waveforms and what problems arise when they are distorted. The art and expertise of spectrum analyzer displays are also given special attention, since this very useful instrument will become much more of a necessity than an expensive luxury as our world moves more and more towards home and business video theatres, along with an enormous expansion of both cable television and two-way radio. The ability to use TDR (time domain reflectometry), as it has been developed over the years, can mean a great deal to those working in long or short cable complexes where precise impedances, coupling, and other critical circumstances must be known within inches to effect necessary design changes or repairs.

This book on oscilloscopes, although containing a good deal of Tektronix and some Hewlett-Packard material for the more explicit applications, is neither one of pure theory nor impractical illustrations. It actually originates from the author's own direct experiences in the intensive use of such equipment, beginning with DC oscilloscopes in the early 1950s to spectrum analyzer work on recent video and two-way radio, beginning in 1976. The results should be highly informative, thoroughly useful, and wholly factual.

Stan Prentiss

Chapter 1
Oscilloscopes and Their Uses

An oscilloscope, strictly defined, is an instrument that displays electrical impulses on the phosphors of a cathode ray tube. An OSCILLOGRAPH on the other hand, also measures similar instantaneous energy values but records them as a *permanent* record. However, the frequency response difference between the two equipments and their measurement flexibilities is vast, and an oscillograph is normally only used in relatively slow waveform examinations where a graph or record on special chart or recording paper is required for indefinite retention. Conversely, oscilloscopes are noted for high-frequency measurements as well as those in the very low frequencies, and can handle signals from DC (ground or common) through megahertz (MHz) and sometimes gigahertz (GHz).

As you may know, there are many types of oscilloscopes in addition to the standard variety known as basic design and service scopes. The more exotic varieties are usually limited to very specific functions such as storage, spectrum analysis, sampling, and logic. In addition, they're also quite expensive since internal circuits are usually more complex and production quantities considerably less. So if you wish to study a random trace over periods of minutes or hours, want to know if overmodulation is the root of broadcast harmonic problems, need to sample extremely fast signals, or desire to look at 8 or 16 streams (bytes or words) of logic simultaneously, you need one of these special instruments that have price tags somewhere between $2,000 and $15,000. But for single- and dual-trace waveform investigations between DC and 15 MHz or 30 MHz, $850 to $1800 can easily satisfy your requirements.

SCOPE BASICS

An oscilloscope—like a vectorscope, which will be discussed later—is basically a power supply and a cathode ray tube (Fig. 1-1). But to supply sufficient horizontal and vertical deflection for small signals and attenuation for large ones, there must also be vertical and horizontal amplifiers. In older oscilloscopes, these amplifiers had individual gain controls, and when they weren't push-pull, either the right or left horizontal/vertical

4 / Oscilloscopes and Their Uses

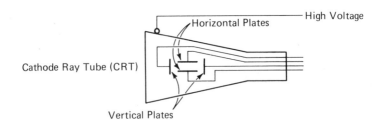

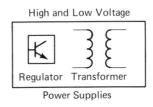

Fig. 1-1. Basic diagram of any oscilloscope-vectorscope.

deflection plate(s) was tied to AC ground, producing rather poor single-ended response. Also, the horizontal deflection sweep oscillator and amplifier response (Fig. 1-2) was roughly marked off in cycles-per-second (cps/sec) so that you would read only a *gross* frequency instead of the more useful time base. In those days—which really weren't too long ago—an oscilloscope of 15% accuracy (vertical amplifiers only) was a treasure to have and hold, and you really didn't care what the horizontal frequency amounted to as long as there was a rough idea of whether you were dealing in Hertz (Hz), kilohertz (kHz), or Megahertz (MHz)—currently defined terms for cycles, thousands of cycles, or millions of cycles.

In the past 10 to 15 years, however, with the advent of semiconductors, widespread use of digital logic and the powerful evolution of computers and microprocessors have forced development of highly accurate oscilloscopes whose vertical and horizontal ranges are much more precise today (within 3%) than were the best analog VTVMs and multimeters of just yesterday. This, in turn, has now prompted the development of digital multimeters whose accuracy varies between 1% and 0.01% (or better), depending on what you're willing to pay. Further, the antiquated idea of measuring AC waveforms with a peak reading meter has now been superseded by true RMS (root-mean-square) reading meters whose DC, AC, ohms, and current functions remain accurate to the same degree.

Oscilloscope progress has taken a similar quantum design leap in that most horizontal sync (and amplifiers) has now been converted from frequency readings into a thoroughly useful time base, producing measure-

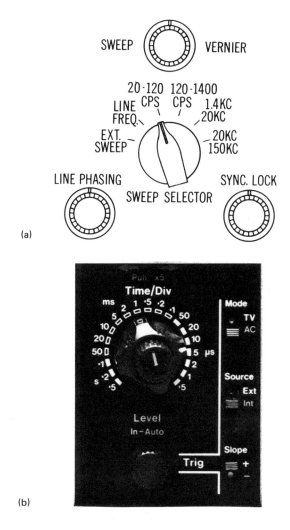

Fig. 1-2. Old style (a) and new style (b) horizontal frequency and time base tuning circuits.

ments that are normally plots of amplitude versus time (Fig. 1–3). This means that one dimension may be a function of the other—usually amplitude A = f(T) Time—offering a two-dimensional look at both transient and recurrent electrical waveshapes. With frequency being the inverse of time, simple division of either quantity into the digit 1 produces the other. For example, if T amounts to 5 milliseconds (msec), then F = $1/5 \times 10^{-3}$, or 200 Hz. Of course there will be much more to say about time and frequency manipulations in the time-base descriptions that fol-

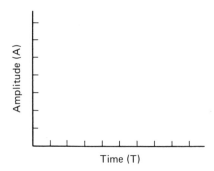

Fig. 1-3. With new time base accurately calibrated from seconds to microseconds (μsec), a useful plot of time vs. amplitude is available for both general as well as critical measurements.

low, since this is of paramount importance in developing the maximum potential of any oscilloscope.

Vertical amplifier progress amounts to a steady evolution of original, selective input, high gain, and low-noise concepts, except that all such circuits are now push-pull and their accuracies well within 3% on better oscilloscopes; while DC offset, drift, and balance—which used to be annoying nuisances on older instruments—have now been reduced so substantially that most external potentiometer adjustments are no longer provided. Standard vertical amplifier ranges for medium frequency scopes ordinarily amount to between 5 and 10 millivolts and 10 or 20 volts-per-division. Usually these amplifier (and voltage divider) limitations produce more than adequate vertical deflection factors for both industrial uses and servicing. Remember, however, that 10:1 low-capacitor probes change a sensitivity of 10 millivolts to 100 millivolts but extend the upper 20 volts-per-division limit to 200 V/div.

You will also discover that the newer and better scopes have well-regulated low- and high-voltage power supplies, and do not deregulate as the older models would with subsequent loss of trace brightness and accuracy. Actually, any worthwhile oscilloscope should maintain vertical-horizontal calibration and beam spot size from 105 to 125 volts at room temperature without noticeable variation. If it doesn't, then it's not worth whatever price the manufacturer is asking. Evaluate *before* you buy! Pause to reconsider whether a simple VARIAC test from 110 to 130 volts line input defeats the instrument's basic specifications.

VERTICAL AMPLIFIERS

Now that you have an overview of general oscilloscope progress and many of the improved features, let's turn to what each usable part of the scope can do for you. Our initial topic, naturally, must be vertical amplifiers, and there are two sets of these for each dual-trace oscilloscope.

Vertical amplifiers will display the amplitudes (height) and shapes (configurations) of all signals delivered to their individual inputs. Dual trace oscilloscopes—those that time-share both A and B traces (channels)—will permit viewing any pair of inputs, either simultaneously at low frequencies in the chopped mode or, alternately, at high frequencies, with first one trace visible and then the other. In this way, special cathode ray tubes are not required, but the advantage of having two traces is retained. There are, of course, split beam, dual beam, and 4-trace oscilloscopes, but these need additional cathode ray tube deflection plates, plus other oscilloscope circuitry, and are considerably more expensive. So here we'll remain with conventional dual-trace types that have great utility for all general and many specialized applications.

As illustrated in Fig. 1-4, at low frequencies the incoming sinewaves are segmented and applied in pieces to an electronic switch, and above a few hundred hertz are delivered at the end of each sweep to the same switch and are seen as individual traces. However, at higher speeds (say kilohertz) these alternately switched signals fool the eye and seem to appear in the same time domain as separate, in-time traces. But both

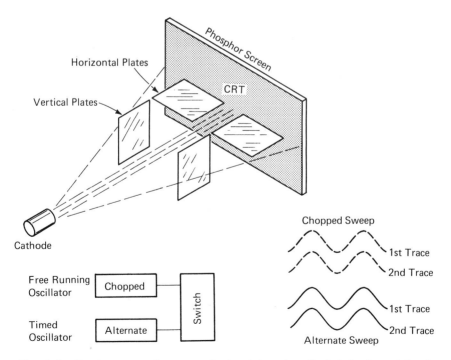

Fig. 1-4. Dual traces using same horizontal and vertical deflection plates in chopped (low frequency) and alternate (high frequency) modes.

signals, even though they may originate from different sources, must ordinarily have identical synchronization or they cannot remain fixed (still) on the face of the cathode ray tube. Occasionally one signal will have twice the repetition rate of the other, and then you may set the scope's sync on the slower one to view both traces. A pair of flip-flops in a ripple counter can produce such a condition where the second trace will divide the first precisely by two. The scope then, could probably sync with sufficient stability so that both rise and fall times and leading/trailing edges of repetitive waveforms could be accurately evaluated.

Rectangular, half-cycle pulses are illustrated in Fig. 1-5, where the difference between the 10% and 90% points in microseconds (usec) or nanoseconds (nsec) is the time it takes for a single pulse to rise to its maximum amplitude or fall to within a few millivolts of DC and rise to Vcc—very close to the supply voltage—depending on transistor leakage and dynamic or resistive load. Here, of course, you're dealing with an ON and OFF switch. In analog waveforms that contain intelligence, there are seldom conditions of either being totally off or driven completely into saturation since one or the other may well produce distortion, especially in direct-coupled amplifiers.

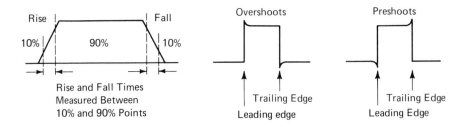

Fig. 1-5. Pulse and square wave characteristics. Note difference between overshoots and preshoots.

If Z-axis (intensity) modulation is required for timing, marking, or other waveform checks and additions, a 20-volt (or higher) input signal into the cathode of the picture tube will intensify portions of the waveform at some frequency of time duration greater or less than that of the original signal, depending on the amount of voltage that needs to be intensified. Lower adjustment of the intensity control will make the lighter portions of the remaining trace quite visible with respect to the darkened portions. But if full trace intensity is maintained, Z-axis modulation may not be seen at all. Should the Z-axis and normal input signals fall out of timing (sync) with respect to one another, then this method of trace analysis is virtually useless because of irregular and often total lack of synchronization; one should at least be some multiple, or submultiple, of the other. AC coupling to the CRT's cathode is through a capacitor, of course.

In general, the aforementioned is what the signal-handling circuits of an oscilloscope do. Applying these principles, however, is another matter, and specific conditions will be discussed just as soon as you've absorbed the meaning and potential of calibrated time base.

TIME BASE

As illustrated in Fig. 1-2, sweep times from 0.5 second to 0.5 microsecond may be selected to view almost any type of incoming waveform. In the nonautomatic mode, a sample of voltage entering the vertical amplifiers is shaped, increased in gain, and delivered to the oscilloscope's horizontal section to trigger synchronizing circuits whose basic frequency is set up by manual adjustment of the sweep timing. In the Auto Mode, a low-frequency oscillator permits continuous trace viewing, and when a signal is received, the time base circuits are triggered (just as they are in the nonautomatic mode) and, upon proper time base adjustment, sync the incoming signal.

Of course, we're going to presume this incoming voltage, whether digital or analog, is at least somewhat repetitive so that several samples will be available to adequately trigger the time base. Should the signal be either random or one-of-a-kind, then *external* sync coupled through the horizontal input connector on the front panel is advisable. Otherwise, you may never see such sporadic voltage as it passes rapidly through the vertical amplifiers. A positive and/or negative level control also aids in triggering on whatever waveform *slope* you wish to start. Among the 59.94 Hz vertical and 15.734 kHz horizontal television sync frequencies, special time constants have been added in the + and − slope positions that will assure further aid for automatic triggering. Invert these two repetition rates, and using a handy calculator, and you will discover that the lower frequency turns out to be 16.68 milliseconds—the time of one field of vertical trace and retrace—and the upper frequency amounts to 63.556 microseconds—the duration of one horizontal trace-retrace line. Comfortable viewing time base settings for these respective frequencies and their reciprocal times amounts to 2 - 5 milliseconds for vertical and 10 - 20 microseconds for horizontal, which amounts to a ratio of 1:3 to 1:6, depending on the number of cycles of each waveform you'd like to see.

Note that nothing has been said about vernier (fine) adjustment of either vertical or horizontal inputs. In their detent (or locked) positions, these verniers ensure that both horizontal and vertical time base and amplifiers are calibrated for maximum accuracy. There are few, if any, conditions where verniers must or should be moved out of their "calibrate" positions. If you really want to discover how much an individual knows about an oscilloscope, check the AC-DC switches on his oscilloscope's vertical amplifiers. If both are in the AC position, he usually doesn't know much. But should vertical and time base verniers rest in their "uncalibrate" positions, then the person using this oscilloscope knows virtually nothing. In other words he/she isn't using the scope as intended, but simply as a general waveform display, which really means next to nothing without meaningful measurements.

In summary, samples of incoming waveforms trigger the better oscilloscopes on either positive or negative slopes of any incoming voltage, with the oscillator of the time base set to a specific time generally coincident with that of signal frequency. Since most signals are at least somewhat repetitive, you have a time-amplitude (Fig.1-3) relationship that may be translated into frequency-cycle by inverting the time of a single cycle and dividing it into one: $F = 1/T$ . . . Frequency equals one over time. As a further example, if you wished to know what time base setting you'd need to view a standard color TV burst (sync) frequency of 3.579545 MHz, your rule of thumb would tell you that somewhere between 500 and 200 nanoseconds (10^{-9} sec.) would be ideal. If used with reason, most of the simpler waveform measurements and their time-to-frequency conversions are rather easy, and we'll do more of these as the indoctrination progresses. First, however, the subject of probes must be introduced and discussed, before the entire oscilloscope measuring system can be adequately covered.

PROBES

There are four varieties of passive (and sometimes active) probes that should be used with the service-type ocscilloscopes (Fig. 1-6). One of these is a magnetic wraparound for current measurements, and the other is a Field Effect Transistor probe used as a wideband demodulator. The latter hasn't yet been placed on the market, although it does, indeed, exist. The other two, of course, are the 10:1 low capacitance and resistive

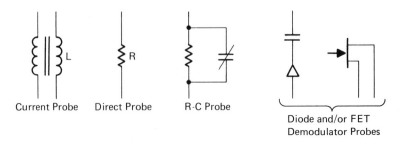

Fig. 1-6. The four types of probes ordinarily used with electronics design and service type oscilloscopes.

isolation direct probes. These are often combined in a single probe body having an adjustable head. Unfortunately, direct probes are used very little anymore since they are relatively low impedance and, with the scope's nominal 1 megohm (Z) to ground (common), tend to load many circuits they might otherwise accurately measure. The attenuating 10:1 low capacitance probe has now largely taken over for all but the detection of RF or IF modulated information in both low and higher voltage investigations. Although this 10:1 probe reduces oscilloscope verticle amplifier sensitivity by a factor of 10, its adjustable capacitance does compensate for low-frequency distortion, while the added 9-megohm series impedance gives the oscilloscope a total of 10 megohms shunted by only a few picofarads, offering very little loading to the measured circuit.

Direct probes, especially in tube equipments, are hardly more than conduits for some rather high impedance circuit to the vertical amplifiers with a median value of about 10K ohms in series with the shielded probe conductor and cable to prevent oscillations. Simplified schematics of the four probes are shown in (Fig. 1-6). The FET and diode demodulator probes are both envelope detectors since they remove either the positive or negative half of the signal, depending on polarity. Of course, this detection is half wave, but should such demodulators be upgraded and advance eventually to become full wave devices, then they would probably be called doubly-balanced, synchronous detectors. Unfortunately, such devices (in oscilloscope probes) aren't yet on the market either. So the envelope diode detector with its usually attendant high-frequency rejection faults is all that's now available.

Since the LC passive 10:1 probe is of prime concern when used with oscilloscopes of 50 MHz bandpass (or less), a closer look at this unit can be quite useful and will substantially demonstrate why probes designed for a specific oscilloscope system should not be used with any scope that's available. As an example, let's say that probe rise time is 10 nanoseconds, and that of a particular oscilloscope is 24 nanoseconds. The equation for the complete system rise time amounts to:

$$\begin{aligned} \text{System } t_r^2 &= t_r^2 \text{ (probe)} + t_r^2 \text{ ('scope)} \\ &= (10 \times 10^{-9})^2 + (24 \times 10^{-9})^2 \\ t_r &= \sqrt{(100 + 576)^{-18}} = \sqrt{676 \times 10^{-18}} = 26 \times 10^{-9} \end{aligned}$$

So, System t_r = 26 nanoseconds. Had a probe with the same ns input rise time of the oscilloscope been used, for instance, the system rise time would amount to 33.94 ns, and that's a substantial increase (and also the usual scope bandwidth reduction) if very sharp rise and fall times of external circuits are to be measured. To be on the safe side, double the system rise time if you want measurement results to be reasonably accurate and free from limiting capacitative shunts.

Fortunately or unfortunately, there's considerably more to probes than simple rise time figures. Look at Fig. 1-7, as an example, and note what often happens to probes with 3.5-foot, then 6-foot cables. You could lose two or more megahertz by just the addition of several feet of shielded wire and its added inductance, since any 10:1 probe consists of an RLC circuit that's always frequency-sensitive. Fig. 1-7, as you may have suspected, is an example rather than an absolute derating curve for all LC probes. We're simply showing that 600 volts peak-to-peak, or any combination of DC and peak-to-peak voltages that add up to 600 volts, can exceed the power handling ability of many LC probes. At low frequencies, of course, this is not really true. In DC measurements, for instance, voltages around 2000 volts can be measured without harming either the scope or its probe. At 60 Hz, the 1600 V p-p vacuum tube verticle TV output voltage may be tested for any reasonable interval without harm, even though the DC potential plus the peak-to-peak waveform will combine and often add up to over 2000 volts. When pulse repetition rates extend into kilohertz and megahertz, however, you'd better pay attention to the 600-volt overall limits, since small probe and scope front-end resistors and capacitors do most assuredly heat up and burn. So high-frequency signals are the real bugaboo of both oscilloscope and voltmeter measurements, and in considering either type of instruments, all AC characteristics need careful examination.

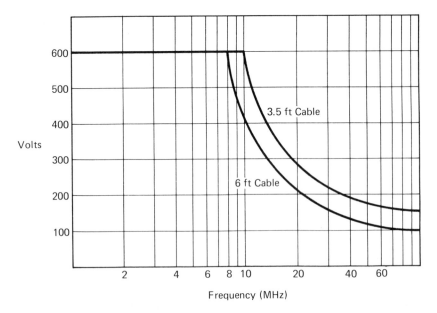

Fig. 1-7. Derating curves for a typical 10:1 low capacitance passive oscilloscope probe with different cable lengths.

CALIBRATION AND AC-DC MEASUREMENTS

There are several other considerations worth exploring as we complete the highly important subject of probes, and these are methods of calibration and AC-DC measurements.

Among vertical amplifiers two methods are equally acceptable, with one very important exception: AC should be used whenever scope front-end capacitance is checked and LC probes are adjusted for best compensation. Squared tops and sides of rectangular (square) waves are needed to be sure no side glitches, droops, or overshoots are present signifying high- or low-frequency loss. Otherwise, a fairly constant DC voltage source such as an accurate digital voltmeter will calibrate your vertical amplifiers very nicely—even to within tenths of 1%, if they can take it. The scope's own ordinary AC calibrate source is normally no more accurate than 2%, which is entirely sufficient for most vertical calibrations. Horizontal time base settings, however, should be highly precise and done with either a special time-mark generator, or if your electronic counter is highly accurate and has a period measurement function, than an AC sinewave source and counter may be used instead. Tektronix time-mark, however, will probably be easier since it can be *dialed* for percentage of inaccuracy and is just a bit more professional.

Additionally, if all calibrations are made with LC probes attached to the oscilloscope, the entire system will be calibrated equally rather than the initially separate adjustment of probes and scope. It's worth remembering that your accuracy is only as good as the system's most *inaccurate* link! And this could well be your probes.

Now for AC-DC measurements, since this subject becomes a little hairy with some people who have been accustomed only to AC terminology. In older oscilloscopes there used to be no DC coupling, and most measurements were for sinusoidal waveshapes only (see Fig. 1-8). As you see, there are both positive and negative half alternations about a zero reference; so, naturally, the figure is called "sinusoidal." This is the type of intelligence of which most analog information is alleged to consist—but

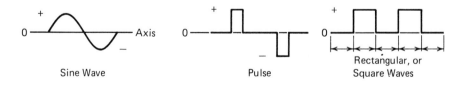

Fig. 1-8. Sine waves, pulses, and rectangular or square waves. Positive and negative pulses could present audio, believe it or not!

really it doesn't. For instance, sound is made up of many pulses, which are often unidirectional and either above or below the zero axis. However, each pulse is seen as wholly complete, being neither immediately joined to another, nor in continuous arc or rotation. So instead of having a peak-to-peak measurement, there is a zero-peak measurement. The same time is true of the following two rectangular pulses whose on and off times are equal (50% duty cycle). Consequently they are called square waves. Again, a zero-peak reading is what your oscilloscope shows, so the term peak-to-peak really has very little meaning in the real world of emerging pulse techniques, sound, composite video, and logic. Therefore, all oscilloscope vertical amplifier measurements should simply be termed "amplitude" and allowed to remain generalized to cover much more than simple sinewave applications.

You might also question if "zero" could not be set at some level other than actual DC, say an offset, and you could be right—set it wherever you want. Usually, however, zero (0) does mean common, ground circuit or system reference, and a pulse rising or hanging from this reference has a starting point of DC. Of course, the pulse (or sinewave) could be "riding" above or below DC, depending on adjusted bias levels; but ordinarily DC is simply the common circuit point and nothing more.

Regardless, the number of divisions a voltage rides or is suspended from some common reference determines not only its amplitude, but, also the DC voltage on which it is carried. For instance in Fig. 1-9, the AC voltage at 5 volts-per division (5V/div.) amounts to an amplitude of 10 volts and appears to be riding on a DC voltage of 5 volts. The latter may

Fig. 1-9. Reading AC and DC voltages simultaneously.

or may not be true, and to find out you'll have to switch from DC to AC amplifiers. Then, of course, the number of divisions the waveform rises or falls determines the actual DC potential on which it is superimposed.

With the foregoing as a fair introduction to using a modern oscilloscope, let's see what these waveform analyzers can actually do. We'll turn now to real, live applications and explain each process in sequence and detail as the discussion continues.

OSCILLOSCOPE APPLICATIONS—LIVE

A sine wave, of course, is always the first example since this probably represents the original conception of an electric generator or motor output. However, with Channel B of the oscilloscope, we'll also simulate a common (DC) base line to illustrate the initial usage of both AC and DC amplifiers in addition to time base.

With time base set at 5 microseconds (usec) per division and Channel A amplitude to 2 V/div. (Fig. 1-10), you see 10 cycles of sinewaves, which are almost precisely superimposed on the 10 vertical graticule bars and indicate horizontal frequency. Here, the sine wave generator's dial is set to 2.1 × 100 kHz, or 210 kHz. Now either the generator is slightly in error or the scope is, because frequency at this time setting should amount to:

$$F = 1/T = 1/5 \times 10^{-6} = 0.2 \times 10^6 = 200 \text{ kHz}$$

Each cycle should be right on the money. We can more or less assume that a sinewave oscillator whose initial alternations are symmetrical and relatively spaced will continue to ring true for a reasonable period. Therefore, looking between the 2nd and 9th graticule lines, you see only little deviation from sinewaves falling precisely on each graticule marking. Since graticule divisions are further marked off in tenths along the center horizontal line, any variations within each vertical line can be roughly translated into 2,4,6,8, etc., tenths. Since the maximum deviation for *any*

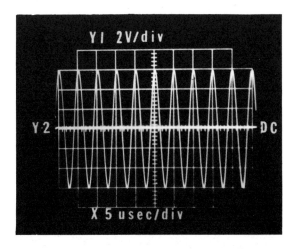

Fig. 1-10. Ordinary oscillator sinewaves can easily be used for oscilloscope time base calibration.

of the off-graticule sinewaves amounts to no more than 0.1, then 0.1/5 × 100 = 2%. That is pretty close for now until some more accurate means of identifying actual calibration is available. If you wish proof positive, we happen to know that the maximum deviation (error) of this particular time base is no more than 1.5% at any point across its measured range; so the 2% estimate is probably more like 1% actual, making an instrument time-mark calibration check wholly unnecessary (at least for this time base setting).

The Y amplifier vertical position supplies equivalent information to the X time base since the Y1 amplitude is selected by the vertical deflection attenuators at 2 volts-per-division (2V/div.). Consequently, the amplitude (height) of the entire waveform amounts to 12 volts, with the DC reference (represented by the Y2 trace) precisely at its center. In all probability, then, this waveform is AC-coupled from the signal generator or circuit in which it's working and has no DC offset of any description. Therefore, you're reading both AC and DC parameters at a single glance and can go on to the next signal immediately (if this one meets specs) without further investigation.

But what if there are DC offsets? These are just as easily read by an extension of the same methods as before. However, you will have to switch from DC to AC amplifiers for a true DC reading. As an example, look at Fig. 1-11. The oscilloscope's vertical amplifier remains at 2V/div., but the time base has been changed to 10 usec/div., and the overall frequency has decreased by about a factor of 10 (0.1). You will also note that waveform amplitude is now 7.6 volts and that DC reference is 2 volts above the most negative swing of the sine wave; so, 5.6 volts of the sine

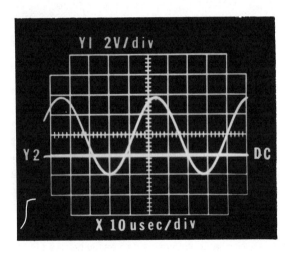

Fig. 1-11. A sinusoidal voltage with DC reference above lowest peak.

wave is positive and 2 volts negative. And how about the new frequency?

Since one cycle from lower peak tip 1 to lower peak tip 2 amounts to 4.8 divisions:

F = 1/T = 1/4.8 × 10 × 10^{-6} = 1/48 × 10^{-6} = 0.0208 × 10^6

Then F = 0.0208 × 10^6, and this reduces to 20.8 kHz. (Dial setting on generator is 21 kHz.) Is this close enough?

Now, let's say that we switched from DC to AC amplifiers and the waveform *fell* two divisions, or 4 volts. What would our DC voltage really be? We could say the voltage was riding on 4 volts DC, couldn't we? Had the sine wave moved positively by the same amount, then the DC voltage would have been negative. Remember, we're only using Y2 to illustrate some general DC potential, not necessarily the exact one that will be developed in each example. Of course, when you have Y2 to look at waveforms simultaneously with Y1, it will be necessary to switch to GND every now and then in case you forget your DC reference.

The preliminaries are pretty well over now, as you can see by Fig. 1-12. Trace No. 2 (Y2), is now turned on, and instead of a simple sine wave, the double trace exposure reveals both sine and square waves, each with different DC references. Furthermore, the time base is now 1 millisecond (1 msec/div.), although the amplitude of 2V/div. for each trace remains the same. But note how the rise and fall times of the rectangular waves are virtually invisible; this is because of the time base setting of 1 millisecond. In comparison with such a slow repetition rate, the very fast up-down sides of the square wave traces are not obviously scanned. As you may have guessed, the sine wave's DC reference remains at its center, but

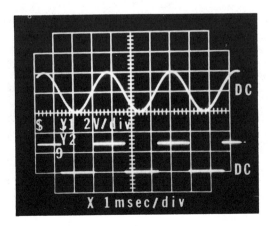

Fig. 1-12. Sine wave and square wave combination. At 1 msec/div. the square wave risetimes are invisible.

that of the square wave below is along the next to last horizontal line from the graticule's bottom, where the pulse is low instead of high. As always in positive logic, pulses swing from DC or millivolts to some greater positive level above DC, equivalent to almost the main supply voltage, less any circuit drops along the way. For instance, if you wanted to know the nominal supply voltage of any square wave generating circuit, just examine the DC-to-Vcc (voltage/collector) swing. Here you could call it 4 volts, since the Y2 amplifier remains set at 2V/div., and the lower voltage advances vertically to about 1.5 divisions.

Another thing you should know about square waves; they will have strange effects on your low capacitance 10:1 probes at both high and low frequencies. In the double exposure picture of Fig. 1-13, you see a perfectly good square wave in the top Y1 trace; but in the superimposed Y2 and Y3 traces, one is much smaller than the other, and one is taller and had distinct overshoots. In Y1, this trace is at high frequency, and any LC probe adjustment simply increases or decreases the height (amplitude) of the trace. But in he Y2-Y3 superposition, the square wave generator frequency has been lowered to 1.3 kHz, and telltale signs of probe misadjustment are obvious. Therefore, with scopes having 400 Hz to 1 kHz internal calibrators, your LC probes are adjusted for minimum overshoot and best rectangular wave response, and this generally takes care of high-frequency adjustments, too. By connecting vertical amplifiers alternately to calibration sources directly through 10:1 probes, you may also check the accuracy of the probe's dividing action. Here, we checked 2 volts against 0.2 volt using an outside square wave source.

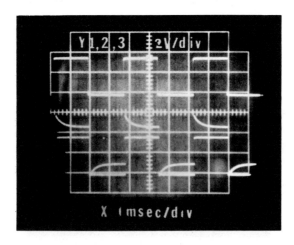

Fig. 1-13. LC probe calibrations are critical at low frequencies. Note overshoots on lower superimposed trace.

Naturally, 1 volt to 100 mV or less (still a 2:1 division) would do the same thing. After the single-channel check, try superimposing Channel A over Channel B and see if they both produce the same amplitudes. Obviously, with only one time base for both traces, horizontal calibration will always be identical.

Now, let's do something with sine waves. Specifically, we reset the oscilloscope's time base to the CHAN B or X position and have a look at the two vertical amplifier phase matches as Channel B is diverted through the Horizontal X amplifier (Fig. 1-14). This of course, is the vectorscope position, which we'll describe next. Meanwhile, take any sine wave generator input and apply to both Y channels of the oscilloscope.

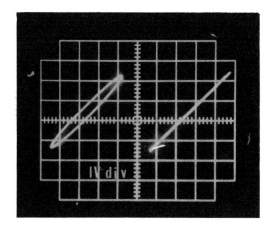

Fig. 1-14. Testing phase relationship between X and Y amplifiers. These are matched up to 30 kHz.

Trace on the right (Fig. 1-14) in this double exposure is shown at 30 kHz, while that on the left amounts to 300 kHz. What you're actually viewing in the way of phase angles amounts to 0°/360° on the right and approximately 30°/330° on the left. So you may say that the Y1 and X oscilloscope amplifiers are matched in this illustration to 30 kHz, and that's really not too bad without special phase matching networks in between. Had the same figures tilted left instead of right, the angles would have been 150°/210° and 180°, respectively.

Now, using *two* signal generators and introducing their individual inputs into Channel A (for the left generator) and Channel B (for the right generator) and then modulating the right generator with a sample of the output from the left generator, different frequencies and external modulation amplitudes can produce both square and round figures, and some in between, that look altogether miraculous. However, it's nothing more than repetitive sinewaves somewhat out of sync that are profoundly affected (the round ones, not the rectangular ones) by modulation into the B generator. Of course, the oscilloscope remains in its vector position throughout Figs. 1-14 and 1-15. In Fig. 1-15, it's interesting to know that 200 kHz was being generated for Channel A and 300 kHz for Channel B.

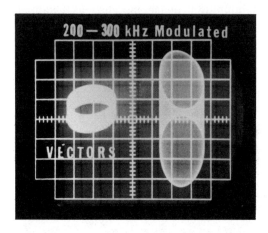

Fig. 1-15. Introducing modulation generates odd Lissajous patterns.

Modulation into the B generator amounted to about 70% from Channel A. Without modulation and if you could sync these pairs of sine waves, horizontal and vertical 2:1, 3:1, 5:1, etc., displays could easily be shown and used as frequency comparison devices at about 5% or 10% accuracy. For much closer tolerances, however, an electronic counter is recommended in all sine wave and pulse counting.

Lissajous oscilloscope patterns (named for Jules-Antoine Lissajous, who died before invention of the oscilloscope in 1880) came into vogue long before present-day counters and their high accuracies were possible. Now, Lissajous images are either emergency tactics or novelties for electronics beginniners or unoriginal instructors. If, indeed, you attach a T receptacle to an oscilloscope input and monitor sine waves or pulses of 1 or 2 volts amplitude (just to be safe) with an electronic counter as they enter the scope, you may read the amplitude, time base, and frequency on the combined instruments at the same time. If the counter measures periods—the inverse of frequency—you may then calibrate the scope's time base with great accuracy simply by dialing whatever frequency/period is suitable for each time base second, millisecond, or microsecond setting. Here, of course, a highly perceptive eyeball helps since confusing parallax always affects accuracy.

VECTORSCOPES

Any vector, as you probably know from secondary school mathematics, is involved both with matrix algebra and polar/rectangular coordinates. Simplified, this all amounts to the magnitude (high, low) of some quantity and its direction. In matrix algebra you have rows and columns to contend with, but in rectangular and polar coordinates there is "operator J"; and this always means phase angle generation, 90° at a time—J^1, J^2, J^3, J^4—for the four quadrants in any electronic circle. The end product is not the simple addition, then, of some arithmetic number—1,2,3 = 6—but a considerably more complex approach whose sum or difference (we'll exclude multiplication and division) is heavily influenced both by waveform content and phase angle. For example, the absolute value of any complex number is often represented by $|a+jB|$ (the magnitude). By squaring the two quantities and taking the square root of the sum –

$$\sqrt{a^2 + jb^2}$$

– their combined value amounts to a real number that is either positive or zero. In electronics, an impedance vector Z that has capacitance and inductance associated becomes:

$$Z = \sqrt{R^2 + (X_L - X_C)^2}, \sqrt{R^2 + X_L^2}, \text{ or } \sqrt{R^2 + X_C^2}$$

And on occasions where X_C is larger than X_L, then X_L is subtracted from X_C. The direction of the vector always points towards the greater reactive quantity. As an example, if our next equation values amounted to R = 4, X_L = 20, and X_C = 12:

Then $\quad\quad\quad Z = \sqrt{4^2 + (20 - 12)^2} = \sqrt{16 + 64}$
And $\quad\quad\quad Z = \sqrt{80}\quad$ or $\quad\sqrt{8.944}$

With reactive inductance X_L being larger, the phase angle would be positive and amount to the arc-tangent of X/R = 8/4 = 63.43°. So the entire vector in Polar coordinates amounts to 8.944 $\underline{/63.43°}$, and that tells you the resultant magnitude of Z as well as the vector's going direction. A graph of this vector is shown in Fig. 1-16. Note that reactive components X_L and X_C are laid off vertically (the inductance upwards and the capacitance downwards), while pure resistance remains at right angles to the others. Operator J simply turns $X_L - X_C$ 90 degrees away from R, but in opposite directions. Consequently, it's impossible for quantities with changing phase angles to be added, subtracted, divided, or multiplied by any other method than vectorially. Therefore, of course, the *Vectorscope*.

Verifying Results

If you'd like to verify these results and pursue vector advantages just a little further, Z can also be checked by its resistance and the cosine of theta (θ):

$$Z = R/\cos\theta = 4/\cos 63.43 = 4/.4473 = 8.94$$

This validates our previous finding. You will also remember from Ohm's law that $E = IR = IZ$. Therefore,

$$E/I = \sqrt{R^2 + X_L^2} \text{ or } \sqrt{R^2 - X_C^2}$$

And from here, either the maximum voltage or maximum current in the circuit can be calculated. In addition, since power is voltage times current in DC and RMS considerations, power can easily be calculated also:

$$P = E \times I, \text{ or } E^2/R, \text{ or } I^2R$$

For those who may have forgotten, however, E is the maximum voltage and I the maximum current. But to find instantaneous values you'll have to use ωt ($2\pi ft$) in the equation:

$$e = E_{max}. \sin \omega t$$
$$i = I_{max}. \sin (\omega t - \theta)$$

where θ is 90°, because across an inductance, current lags voltage by 90°. In a capacitative circuit, $-\theta$ becomes $+\theta$, so that

$$I = I_{max}. \sin (\omega t + \theta)$$

since current now leads voltage. In the actual example of Fig. 1-16, however, the direction turned out to be inductive (with a positive phase angle), so an inductive θ would be the correct phase angle in the equation for instantaneous current.

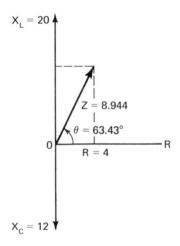

Fig. 1-16. Illustration of RLC vector addition to find resulting impedance Z.

In Color TV

Probably the best way to describe a popular use for the vectorscope is in color television. For here it is indispensable in chroma and subcarrier alignment and in diagnosing troubles among the bandpass amplifiers, the 3.579545 MHz subcarrier oscillator, and separate chroma or combined luminance outputs. In these applications a vectorscope, in addition to a good, clean color bar generator, is much more useful than a sweep generator and is phase-accurate as well. So one can say that the color bar generator is dynamic, while a sweep generator is static. Further, when color bar signals pass through the UHF-VHF tuners and video intermediate frequency amplifiers to the chroma processor, any tilt induced among circuits preceding the chroma processor becomes compensated for in the overall chroma alignment, making color reproduction as true as it can be for any particular receiver. A single dual-purpose diagram can readily show why (see Fig. 1-17).

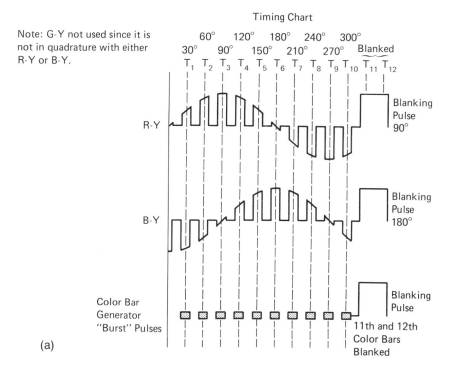

Fig. 1-17. Color timing chart (a) and vector wheel (b) on next page show sequence of 10 gaited rainbow colors as they appear on TV receiver's cathode ray tube. Vectors align as well at troubleshoot.

24 / Oscilloscopes and Their Uses

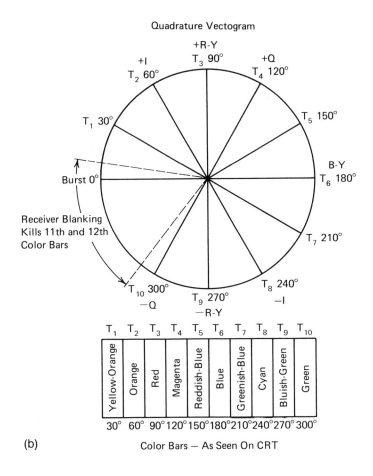

(b)

Note that the R (red) − (minus) L (luminance) and B−Y (blue minus luminance) are at quadrature when taken from color receiver outputs into the picture tube. In the newer receivers with good black level clamping, just turning down the brightness will often suffice; but in older sets, where luminance is combined in either the RGB output amplifiers or in the demodulator, most luminance will have to be shunted to ground through an 80 uF capacitor somewhere in the luminance circuits shortly after the delay line. This is usually done experimentally until the sharpest R−Y, B−Y trace is vectorially produced. If you don't remove luminance, the combination of color and video will both distort and smear the R−Y, B−Y information, making it virtually unusable. Old tube and hybrid sets whose picture tubes matrixed luminance and chroma offer no problem since color goes only to the grids of the cathode ray tube.

So connect the Channel A low capacitance probe at 50 V-div. (for hybrid and vacuum tube receivers) and 20 V/div. (for solid state sets) to the R−Y receiver output, and the Channel B low capacitance probe to the B−Y output (same voltage levels) using AC coupling. Switch the scope to its vector position, and read the results on the cathode ray tube. (See Fig. 1-18, showing both RB−Y waveforms along with a double-exposure vector in the center.)

These waveforms include a reasonably symmetrical pattern—10 spokes (or petals) at 30° intervals between 30° and 300°, with the 11th and 12th bars removed by receiver blanking and flyback interval, which amounts to between 11 and 12 microseconds on most receivers. The 3rd (R−Y) and 6th (B−Y) petals denote the angle of demodulation (almost precisely 90°), and good bandpass alignment (for a very poor alignment see Fig. 1-19—an actual misalignment) is apparent when the 3rd bar has fastest

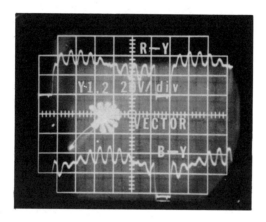

Fig. 1-18. R-Y and B-Y color bars and a vector pattern in the center.

Fig. 1-19. Gross bandpass transformer tuning.

rise and fall times (for greatest positive swing) and all other bars are shaped as much as possible like the 3rd, with absolutely no petal crossovers. All pattern spokes should be distinct, although on poorer receivers they may not always open completely.

If you find problems, begin with the chroma takeoff coil, progress to the 1st and 2nd (or output) bandpass transformers, and tune the chroma processors for best overall results. Then, with the tint control set to its mechanical center, adjust the tint phase control circuit so that the 3rd bar (normally true) is perpendicular to its horizontal axis. Rotate the tint control to turn the pattern at least 35° on either side of center setting without chroma fallout or excessive pattern distortion. See Fig. 1-20 for actual waveforms. If there is trouble with burst phase, loss of color sync,

Fig. 1-20. A good vector pattern must turn at least 35° right and 35° left without undue distortion if chroma alignment and tint control settings are accurate.

etc., both the burst amplifier transformer and 3.58 MHz oscillator transformers may be turned for maximum color sync control (when tint circuits aren't in parallel with color controls). Otherwise, bandpass transformers and control color sync must be tuned accordingly. Under such circumstances, burst and subcarrier circuits should not be disturbed until the bandpass touch-up is completed. The 3.58 MHz oscillator, of course, also supplies R−Y and B−Y reference sinewaves for the chroma demodulator, and you can look immediately at the 3rd and 6th vector petals to see if references are satisfactory before either tuning or troubleshooting procedures begin.

Troubleshooting

In troubleshooting, first look at the pattern's size. If this is what you see with standard verticle amplifier settings, then go directly to the bandpass amplifiers since their output is totally inadequate (Fig. 1-21).

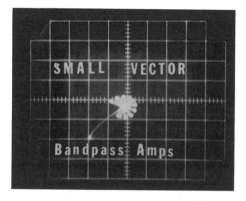

Fig. 1-21. A small vector pattern means bandpass amplifier troubles.

Use the color bar waveform in the chroma circuits and an oscilloscope with a 10:1 LC probe. You'll find a lack of output promptly. When the good vector is spinning, as the one on the right in Fig. 1-22, you know immediately your subcarrier oscillator is completely out of sync. You must now determine if no burst is reaching it or if the oscillator, itself, is defective. Again, signal tracing and adjustments will find your problem.

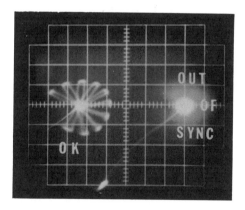

Fig. 1-22. A good vector on the left; oscillation on the right.

The final fault is simulated, but it is real, nonetheless. When one or more (but not all) colors predominate or are missing, the usual difficulty lies directly within the red and blue output amplifiers. As the vectorscope illustrates in Fig. 1-23, either $R-Y$ is down or $B-Y$ is affected. Again, an oscilloscope check should find the cause in short order since you already know where to look. $G-Y$ is not used.

28 / Oscilloscopes and Their Uses

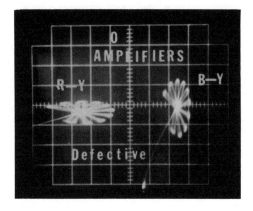

Fig. 1-23. R-Y and B-Y voltages show compression when their respective amplifiers are at fault.

Vectorscopes aren't necessarily useful in everything, but applied judiciously and in their place, they are a powerful television and color monitor diagnostic, alignment, and repair tool. We strongly believe that other equally productive applications will be found as vectors become more widely understood and used.

PROFESSIONAL OSCILLOSCOPES

The difference you might have expected between modern "service" oscilloscopes and what are known as "professional" oscilloscopes probably does not really exist. Both varieties today are triggered sweep, have at least 5 percent vertical and time base accuracies, often include dual trace versions with A + B and A − B cumulative or differential amplifications, and both have AC and DC vertical amplifiers—the magic difference here being little more than a capacitor for AC and a switch shunting the capacitor for DC.

Among other qualities, however, the vertical deflection factors and time base ranges are often quite different. The higher priced professional oscilloscope is usually spec'd at 3 percent accuracy overall (usually this means about 2 percent), will display signals from 2 millivolts per division to either 5 or 10 volts/div. and has time bases adjustable from seconds to low nanoseconds per division. All components, especially semiconductors, are of the highest quality, and every possible safeguard is built in.

Like the better dual vertical amplifiers with delay lines connected so that the leading edge of most or all waveforms is visible, there are also dual time bases with, say A delayed by B, so that small (incremental) portions of any signal can be mixed (regular and delayed times) or displayed across the entire graticule as a portion of some voltage which can be examined in detail at a *lower* time base rate. One especially exotic 7B53A (Fig. 1-24) that's used in the top-of-the-line 7000 Tektronix series has normal sweep, intensified, delayed, and mixed; will trigger up to 100 MHz, with standard sweep rates of from 5 seconds to 50 nanoseconds per division; and has an X10 magnifier that reduces sweep as low as 5 nanoseconds/div. Pushbuttons are also provided for automatic sweep,

7B53A

Fig. 1-24. Super professional regular and delayed sweep time base with pushbutton mode selection.

normal (nonautomatic sweep without baseline), single shot and reset, AC horizontal coupling, AC low-frequency reject, AC high-frequency reject, and DC coupling, as well as four additional switches for internal or external, line, and external input + 10, which probably means an additional 10X amplification. You will also find delayed or direct trigger inputs, trigger levels, and delay time multiplier control that's used with the second time base for incremental voltage selection. Finally, there is a sweep calibration potentiometer slot on the front of the 7B53 that will adjust gross calibration for all ranges, which can be easily and accurately done with a good electronic counter and sinewave generator using, say, the 100 Hz, 10000 Hz, 1MHz, and 100MHz ranges. Simply divide 1×10^{-2}, 1×10^{-3}, 1×10^{-6}, and 1×10^{-8} into 1 for correct frequencies. Then use the first and ninth vertical graticule markings (Fig. 1-25) references and adjust for best overall response. This, or any other time base, may be calibrated as accurately (probably more so) by this method as with much more expensive equipment. After all, many good counters are accurate to

30 / Oscilloscopes and Their Uses

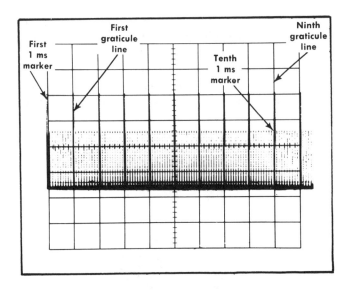

Fig. 1-25. Calibrating time bases with spikes or sine wave markers.

1 part/million, and that's considerably better than any oscilloscope yet built.

Of course, all new oscilloscopes haven't plug-in modules, and many so-called portables meet most specifications of their more exotic looking modular big brothers. For instances, Tektronix (and others) have relatively lightweight 100 and 200 MHz scopes that do the job with fixed vertical deflection and time base units, although without the flexibility that various plug-ins offer. Conversely, portability is a considerable asset, and with other factors being relatively equal, the price differential between exotic and standards is more than considerable. So you may want to weigh carefully all your requirements before making that large financial commitment. Naturally, the source for your instrument will (or should) become a considerable factor in making the final decision. For straightforward laboratory-type scopes without storage, sampling, TDR, etc., there are at least six manufacturers and assemblers in the U.S. who qualify. Many of the technical specifications will be the same, so deciding factors are often price, repair facilities, parts, availability, and manufacturer's reputation. Don't, however, overlook the magnitude of voltages to be measured. Remember that a 10X low capacitance probe and an 8 × 10 division graticule in a 10V/div. scope can handle 10 × 8 × 10, or 800 volts total on the entire face of the cathode ray tube. Obviously, if you're looking at 1000- or 1200-volt pulses occasionally, a 20-volt/div. scope (and not a 10V/div. instrument) will have to be used. Some engineer-technicians miss this fine point, and oscilloscope makers will also tell you that 600 to 800 volts is all their scopes will handle safely. But this is not true when looking at either DC or slow AC voltages. Most 20V/div. scopes will accept signals in these two categories up to 1500 volts without damage, at least for short periods. But fast high-peak voltages, even in kilohertz, can readily burn 10× probes with regularity. In these instances, however, the scope itself is usually spared.

There are also excellent portable scopes available from this manufacturer and others with outstanding features such as the ability to view both vertical channels and their sums or differences and an external trigger trace, all at the same time. This new wrinkle allows accurate time comparisons between the incoming signal and scope trigger at any sweep speed which, in Tektronix' 100 MHz 465B, has now been increased to 2 nanoseconds per division. In addition, there is alternate sweep switching to allow viewing of both A and B delayed sweeps simultaneously at full CRT screen widths, plus a separate intensity control for the B sweep and trace separation control (Fig. 1-26) for both.

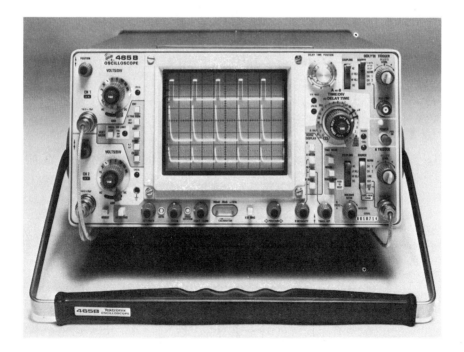

Fig. 1-26. Portable scopes, such as the 465B, have interesting innovations and improvements with faster sweeps and trigger viewing as the third trace.

Uses for these better oscilloscopes also differ from their slower time base and lesser bandwidth brothers. Dual-trace digital signals with very fast rise and fall times that may or may not have glitches or jitter (Fig. 1-27) are the targets of these fast scopes. As Fig. 1-28 shows, rise (and fall) times are measured between the 10% and 90% points. When the bit rate is extremely fast—as in T^2L or emitter coupled logic—scope bandpass and triggering have to approach 100 MHz and nanoseconds even to follow the logic itself, let alone what's happening at the leading and trailing edges. Faster sweep speeds are also the reason oscilloscopes must have 10 kilovolts or more accelerating voltage rather than excessive cathode ray tube drives to see rapid waveform configurations. Up to 4 kV

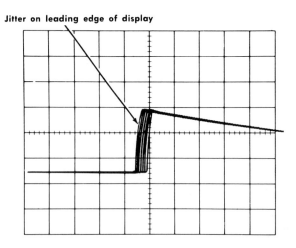

Fig. 1-27. Better scopes can see leading edge jitter because of much faster sweep speeds and excellent stability.

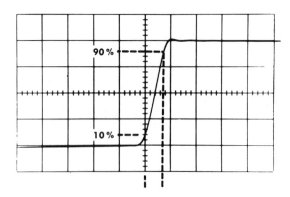

Fig. 1-28. Measure rise and fall times between 10% and 90%.

Professional Oscilloscopes / 33

is good for little more than 10 to 15 MHz. Beyond this bandpass, at 3 dB down, 10 kV is usually a necessity, considering 1980's state of the art.

With the advent of more and more integrated circuits, masking, highly controlled substrate layout, and low operating voltages pretty well eliminate jitter from pure IC sources. But where there are discrete transistor coupling networks having resistors and capacitors as time-constant elements, jitter often plays an important role in circuit operations and should be carefully evaluated. An ambiguous voltage on the trigger surface of any waveform usually results from other errors somewhere further into the system. And there are often times when these networks must be isolated and investigated alone to find both undesirable design and component failure troubles. Let your oscilloscope, then, become an electronic eye on which you may always depend.

One final note before ending the chapter: if you wish to sync internally on low-repetition-rate voltages, be sure your scope can be switched manually into the chopped mode; otherwise, the scope will probably have difficulty holding any trace stationary around 60 Hz.

Fig. 1-29. Sencore markets a new 60 MHz, dual trace, microprocessor-controlled scope having 5mV to 20 vertical V/div deflection factors and time base to 100 nsec.

Chapter 2
Spectrum Analyzers

The objective of ordinary oscilloscopes and highly complex spectrum analyzers are quite similar. Both are concerned with the *amplitude* of any signal; but for other responses, an oscilloscope measures time domain and the spectrum analyzer measures frequency domain. Therefore, a spectrum analyzer will display either complex or basic voltages in terms of amplitude and frequency, the latter being the inverse of time. (Fig. 2-1).

The spectrum analyzer, however, does a great deal more than just recognize some simple response. With extended ranges it can identify a very large number of frequencies simultaneously or individually, and with the aid of a tunable filter, is able to "read" the resolution of separate signals so their bandwidth characteristics may be accurately determined. Further, with storage or flood-gun capabilities, such images are held motionless on its cathode ray tube for minutes or hours while the operator either studies or photographs every detail. Consequently, wide-band analyzers with frequency ranges from Hz or kHz to MHz and GHz, and resolutions between Hz and MHz, are very useful instruments indeed, especially for broadcast video, radio RF and CATV, as well as internal information, in virtually all transmitter-receivers (transceivers), whether they be voice, video, or other types of communications. The spectrum such signals occupy, in addition to their specific characteristics, are of vital interest to national and international regulatory bodies and of equal benefit to manufacturers and users alike. With the increasing use of all spectra, good spectrum analyzers are today considered necessities rather than luxuries.

Spectrum analyzers come in various packages from several different manufacturers. The less flexible variety may be portable and/or relatively narrow band instruments having somewhat equivalent prices. Others, with a variety of options and 80 dB dynamic ranges that even include double-digit gigahertz spans are often microprocessor controlled and even internally digitized for signal storage in memory. With enticing options, such equipment can cost in excess of $30,000 and can even feature bus transfer measurements out of the instrument for "further interaction by a computing controller." In the business of "Signal Analyz-

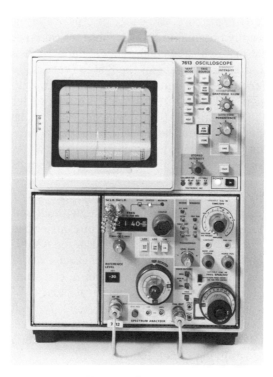

Fig. 2-1. Tektronix 7L12 100 kHz to 1.8 GHz spectrum analyzer plug-in and 7613 oscilloscope mainframe. Blank space on left receives other plug-ins such as preamplifiers. (Courtesy of Tektronix)

ers," as Hewlett-Packard describes them, you can have a lot or a little in spectrum analyzers, depending on the pocketbook. Your author had the pleasure and privilege of working with an 8565A $18,900 H-P analyzer, covering the range of from 10 MHz through 22 GHz, and the results were superb. But unless you're a rather wealthy laboratory or part of some large corporation, access to this type of equipment is a little difficult; consequently, we'll stick to the less exotic varieties of analyzers, knowing that anyone who understands these less expensive units can "graduate" to something more deluxe quite easily when and where the occasion demands.

Spectrum analyzers are not exceptionally difficult to use as soon as you learn their inherent limitations. For instance, the allowable magnitude of input voltages must not be exceeded unless you want a "front end" repair bill that's often huge. Center frequencies and reference marks are necessities in any type of spectrum measurement, and the instrument must have some sort of phase-locked loop arrangement to successfully measure non-drifting, narrow band resolutions. Finally, decibel (dB) inputs from

+ dBm to −110 dBm or more, can easily be cross-referenced to millivolts and volts via readily obtainable scales. These, of course, are all tallied in the final calculations, which also normally include 10 dB steps on the log scale for each horizontal division of the 8-step (vertical) graticule. If, for instance, the fundamental input appeared within graticule limits at −50 dBm, and harmonics showed at 10 dBm steps down, such harmonics could then be said to fall a total of 80 dBm down, or −30 dBm, referenced to −50 dBm. Normal input impedances for such analyzers is usually 50 ohms, although 75-ohm terminations are usually available as options as well as some 10X isolation probes and 5.9 dB 75 to 50-ohm BNC impedance matchers.

Actually, any worthwhile spectrum analyzer is a very sensitive swept-band receiver that is frequency-stable, without spurious responses of its own, and displays the absolute value of Fourier components in any incoming waveform, whether CW, pulsed, FM, or AM. Thereafter, basic signal amplitude or frequency modulation characteristics, frequency conversion products (mixers), pulsed power, noise, etc., can all be identified and measured for both response and spectral purity. What you will be able to see in detail depends on the analyzer's dynamic range and its resolution of both frequency and amplitude. Good resolution should amount to 1 kHz or less in any of the better instruments, with lower frequency units resolving under 100 Hz.

APPLICATIONS OF SPECTRUM ANALYZERS

Now, with these explanations digested, you should be able to go on smoothly to the various applications of spectrum analyzer uses, which is the prime purpose of this entire chapter.

Amplitude Measurements

Amplitude measurements on any worthwhile spectrum analyzer are produced in either decibels (dB) or volts/division. The former is logarithmic (log), while volts/division (lin) is read as a linear function of the usual eight horizontal graticule lines found on either spectrum analyzers or ordinary service oscilloscopes. In any event, the baseline for either log or lin is referenced to the bottom horizontal line for a calibration point; afterwards, measurements are taken in dB or volts/division steps (see appropriate tables), depending on any attenuation (Reference Level) and subsequent voltage levels. In the Tektronix 7L12/13 series, for instance, the RF reference level, dBM step factor, resolution, and frequency/division are all apparent on the cathode ray tube via time sharing during the blanking interval (Fig. 2-2) that does not interfere with the spectral curves illustrated.

Applications of Spectrum Analyzers / 37

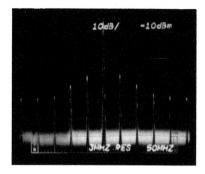

Fig. 2-2. Typical spectrum at 50 MHz/div., 10 dB per division, and −10dB to start, so that initial "markers" off center frequency would read −45 dB (or 45 dB down). The resolution within each calibrate point amounts to 3 MHz. Center marker is an amplitude reference and is not used in spectrum display calculations.

Resolution

Resolution actually defines the ability of an analyzer to separate closely adjacent signals so they can be individually observed. These may be complex or not, but it is the analyzer's job to single out portions of each and display them as typical sinewaves, depending on the instrument's specified resolution. From low Hz to MHz are good and sufficient numbers. Included in the description of resolution is the shape factor, a figure of merit that tells whether the analyzer can show small signals as well as large ones in the same response. As Morris Engelson of Tektronix puts it, "a 10:1 shape factor means that a 60 dB bandwidth is 10X the specified resolution bandwidth." A 4:1 shape factor is excellent (Fig. 2-3).

Bandwidth and resolution, while seemingly equivalent, are not precisely the same because of the analyzer's tunable filter. Bandwidth is actually dependent on amplifiers within the analyzer as well as sweep time/division. Therefore, uncalibrated conditions where trace amplitudes

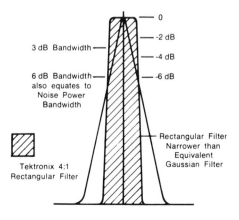

Fig. 2-3. Comparison of the Tektronix exclusive rectangular filter shape factor to the widely used Gaussian filter. (Courtesy of Tektronix)

decrease and resolution increases affect both vertical and resolution calibration. Better instruments either have an "auto" position or a warning indicator when sweep time, bandwidth, and frequency spans overexpand their collective relationships in resolving adjacent signals. Resolution very often is characterized by more than one signal such as opposing skirt merging of two equal sinusoids into a 3 dB down notch. But since ideal situations such as this aren't usually possible, the term *resolution bandwidth* is often given as the spectrum parameter most suitable in describing such measurements. Don't, however, confuse resolution bandwidth with Hz, kHz, or MHz per division. Resolution bandwidth is the inner measurement of the trace, while the external perimeters are simply the number of divisions each waveform occupies and are always identified (by Tektronix, anyway) as the lower right number on the graticule. For single waveform measurements, bandwidth resolution is usually taken at the 6 dB down (voltage) point, which is ordinarily sufficient for either narrow or rectangular waveshapes. As Mr. Engelson so aptly points out, a 4:1 shape factor is essentially rectangular. In actual waveshape measurements, resolution bandwidth needs to be narrow, avoiding all possible noise, but not so narrow as to introduce distortion. Where there is little external and/or internally generated noise, make your measurements at the greatest possible resolution bandwidth, and by all means, use video and kHz front panel filters to remove noise when analyzing critical waveforms. Of course, a 2 dB measure per division will introduce a great deal more noise than the usual 10 dB. So when undertaking analysis of distortions, signal/noise values, or complex waveshapes, do choose the best instrument settings for prime results. In communications work, there is really no other means to fully evaluate the significance of many spectrum conditions other than with a spectrum analyzer. CAUTION: Analyzer inputs will only accept a certain amount of fairly low-level signal. Consult your service manual before applying incoming voltage, and always begin with some input attenuation (say, −30 dBm) when introducing inputs. If external padders are also used, then their dBm values must be added to the analyzer's own attenuator settings when reading out results.

Frequency and Time Bases

These functions, simply the inverse of one another (Fig. 2-4), are fundamentally trigger circuits quite similar to those in conventional oscilloscopes and may be classed as *internal, external, line,* or *free run, single sweep,* and *normal*. Out of the calibrated SPECTRUM position, such a time base can sweep from 10 msec/div. to 1 usec/div. Note that we're talking "sweep" now rather than simply time base, since the purpose of such circuits is simply to present a swept display that can be easily "read" and measured. Along with the 10 kHz, 30 kHz, and video filters, as well as the frequency span per division (which, of course, determines what

Applications of Spectrum Analyzers / 39

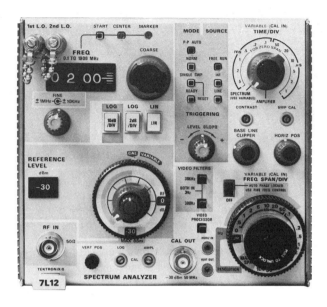

Fig. 2-4. The 7L12 spectrum analyzer plug-in that fits any Tektronix 7000 series mainframe. Center frequency here is set for 200 MHz, reference level at 30 dB down, with time base in the *spectrum* position. The coarse center frequency controls only the higher frequencies, while the small dual fine tuning knob on the left centers all traces in the phase-locked loop mode during lower Hz and kHz investigations.

frequency and resolution you're going to look at) develops f(X)=Y portion of any spectrum readout. There is also a video processor to help with filtering in video situations. Of course, the FREQ SPAN/DIV control on the lower right of the 7L12 series operates strictly in conjunction with the SPECTRUM position on the Time/Div selector. Otherwise, of course, you're looking at time rather than frequency. If triggering is uncertain, a level slope control can help.

In order for the FREQ SPAN/DIV to operate at its best potential, Tektronix has instituted a phase-locked loop function for frequencies less than 0.2 MHz, and with the AUTO button in the PLL position, the instrument goes directly into the PLL mode as it is switched into the yellow area of the dial. Total resolution here, of course, is specified at 300 Hz to 3 MHz (Fig. 2-5 A, B), while the frequency span extends from 100 kHz to 1.8 gigahertz (GHz). As the FREQ SPAN/DIV control is manually rotated, both frequency and resolution conditions are coordinated automatically and displayed digitally on the lighted graticule. When using the spectrum portion, the 7L12 is operated only in the "Right " position for the overall large module control. Left would include a CATV

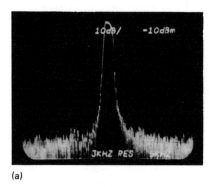 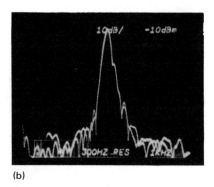

Fig. 2-5. Two photographs of any vertical display "expanded" to show 3 kHz and 300 Hz resolution of identical waveforms. Note the differing measurements at 5 kHz/div., and 1 kHz/div. due to the analyzer's internal filter.

preamplifier, which is used for additional sensitivity—an extra plug-in that is not included in the analyzer's base price. Horizontal positioning and baseline clipper (noise removal) controls are also supplied for further convenience in visualizing the trace.

AM SPECTRUM MEASUREMENTS

You must always realize that a spectrum analyzer is actually translating some time domain representation into what is known as a Fourier spectrum. Credit goes to Baron Jean-Baptiste Fourier, a French mathematician, who discovered that any signal may be the sum of its dc component and an infinite quantity of ac (sine wave) components. One dc output by itself results in a signal of constant amplitude, while an average energy display exhibits only a constant magnitude. Voltages, therefore, are seen in RMS (root-mean-square) representations as positive values. Simple power, consequently, may be calculated in terms of $P=E^2/R$, I^2R, or EXI and derived quite easily if either current or resistance is known. In the end, the analyzer measures only an internal filter output with which it plots some spectral display. Many displays will be investigated, as the discussion becomes both more specific and complex.

AM Modulation

As with any spectrum analyzer, dial or set the indicated carrier frequency, check calibration reference and zero (lin) or dc reference, be sure of sufficient RF attenuation, key the transmitter, and measure the modulation. In the *linear mode*, for instance, the ratio of a sideband to carrier ampitude (Fig. 2-6) could equal 2/8, as shown, which would measure 0.25, or *half* the modulation index. Therefore the percent of AM modulation (with both sidebands) would amount to:

$$\%\text{AM Mod.} = 100 \times 2 \times 0.25 = 50\%$$

AM Spectrum Measurements / 41

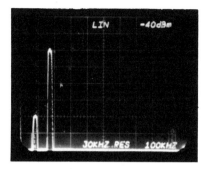

Fig. 2-6. Measuring AM modulation by carrier-to-sideband amplitude in the Linear mode. Resolution and frequency/div. settings are apparent.

The greater the ratio, the larger the modulation, and vice versa. This simple method works whether the analyzer directly "reads" the transmitter via a common connection or picks it up from an on-the-air source such as an antenna.

AM Distortion

This parameter is read very much the same way; but here, you will use the log scale instead of linear. Then simply read the *difference* in dB between the carrier and sidebands. If, as shown, it measures 40 dB, this amounts to a voltage loss of 0.010. So AM distortion becomes:

$$\%AM \text{ distortion} = 100 \times 2 \times 0.010 = 2\%$$

The actual equation that should be used develops as:

$$10 - \frac{dB}{20} \text{ difference}$$

which would become 10^{-2} or $1/10^2$, thus proving our direct reading of 40 dB loss equals 0.010.

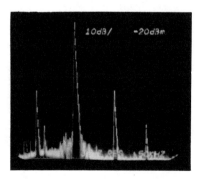

Fig. 2-7. Calculating AM distortion in the Log mode as simply the difference in dB between the carrier and its sidebands. Here we are working at 300 MHz resolution.

Harmonics and Spurs

Harmonics and spurs of course, are read directly on the log scale as so many dB down, and their resolutions and separations from the carrier may be calculated directly from the calibrated graticule. There are no equations here, just simple arithmetic. You may find the addition-subtraction process easier if you move the analyzer's center reference fully left and expand the measurement scale for better accuracy (Fig. 2-8). In addition, if the spectrum display also includes storage, then a nonmoving waveshape can more easily be measured (Fig.2-9). However, since most spectrum analysis consists of repetitive signals, the storage feature isn't mandatory except where there is a nonrepetitive burst of energy followed by other signals that do repeat.

If you're measuring just for general information, the upper and lower sidebands should respond with equal amplitudes. If not, you'll undoubt-

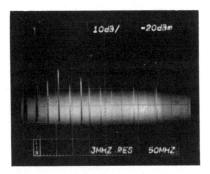

Fig. 2-8. The second vertical spike of voltage from the left is the carrier, and other signals can be regarded as harmonics. The third "harmonic," for instance, would be read as −48dB or 48dB down.

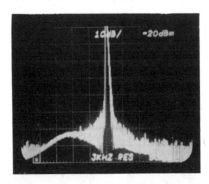

Fig. 2-9. Analyzers with storage features are very important, especially when a single transient or "burst" voltage requires study or analysis.

edly discover a transmitter tuning problem that needs correcting. Successive harmonics, however, will often bear little amplitude relation to one another except that they must meet Federal CB specifications of for instance, at least 60 dB down or better. Otherwise, examinations of the entire collective signal and any interfering frequencies should be considered routine.

Signal/Noise Measurement at RF

Connecting the analyzer to a single RF test loop or vertical whip will quickly give you an indiction of S/N potential. Use the 10 dB/div. and 100 kHz/div. selectors, set transmitter carrier on the top graticule line (if possible), fire the transmitter, engage analyzer's video filter, and measure the dB difference between the carrier and the video filtered display, which, of course, is considerably different from conventional noise. Subtract a 21.2 dB bandwidth correction factor (according to Tektronix) from the difference, and this will be your true S/N ratio (Fig.2-10).

$$S/N = S/N \text{ measured} - 21.1 \text{ dB}$$

For audio, signal generator at microphone input delivers 100 percent audio modulation tone.

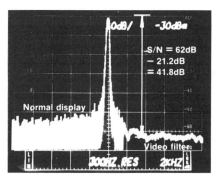
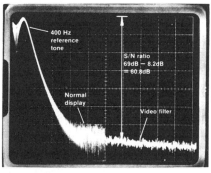

Signal-to-Noise Measurements at **rf**   Signal-to-Noise Measurements at Audio

Fig. 2-10. Signal/noise evaluations of both audio and RF displays, including negative correction factors of 8.2 dB for audio and 21.2 dB for RF. (Courtesy of Tektronix)

Oscillator Distortions

Oscillator distortions become readily apparant when you understand that nonharmonic oscillator products are spurious. Careful filtering, adequate bias, service and adjustments minimize these spurs, which are always evident on any worthwhile analyzer. View your display so the oscillator under investigation isn't loaded.

FM SPECTRUM MEASUREMENTS

Frequency modulation generates sideband frequencies whose amplitudes are a function of both the modulation index and its modulation frequency. As defined, the *modulation index* is the quotient of total carrier deviation divided by its greatest modulating frequency.

Mod. Index = Peak carrier deviation/Peak modulation frequency

As the modulating information increases, so does deviation of the oscillator frequency. It follows that the higher the modulating frequency, the greater the oscillator deviation frequency. The actual number of generated sideband pairs, then, depends on resultant modulation index, which is directly influenced by both deviating and modulating frequencies. Usually, the higher the modulating index, the greater the sidebands. And, as a direct consequence, high modulating frequencies for specific deviation produce less sidebands and greater spacing; but with more sideband pairs, bandwidth emission becomes greater. Transceivers that broadcast modulation frequencies of 10- or 15-kilohertz obviously occupy larger bandwidths than radios having 1- to 2- kilohertz deviations. This may be expressed as a *deviation ratio*:

Deviation Ratio = Max. permissable deviation/Max. mod. frequency

As you can see, the *deviation ratio* differs from the *modulation index* in that it specifies Max. permissible deviations and frequencies rather than Peak carrier deviations and Peak modulation frequencies. In 2-way radio, ± 15 kHz is known as wideband fm and ± 5 kHz as narrow band fm. By limiting both frequency and maximum deviation, fm becomes highly useful to voice communications at sensitivities that are often below one microvolt. For output power in fm stations, this amounts to the effective radiated power (ERP), which may be calculated in three ways:

ERP = (1) Power at ant. × ant. power gain
 = (2) Power Xmtr. − Power line loss × ant. power gain
 = (3) Power at ant × (ant. field gain)^2

Antenna gains, naturally, are calculated with respect to some "standard" antenna which may range from an ordinary half-wavelength dipole, to an isotropic or quarterwave whip. Passive devices, of course, have no actual gain.

FM Modulation

The easiest and most accurate way to attack the measurement of fm modulation when using a spectrum analyzer is via Bessel Null functions. These have been worked out by Tektronix in a Bessel Null Chart shown in Fig. 2-11. This, in turn, is derived from the Bessel Null Functions chart

FM Spectrum Measurements / 45

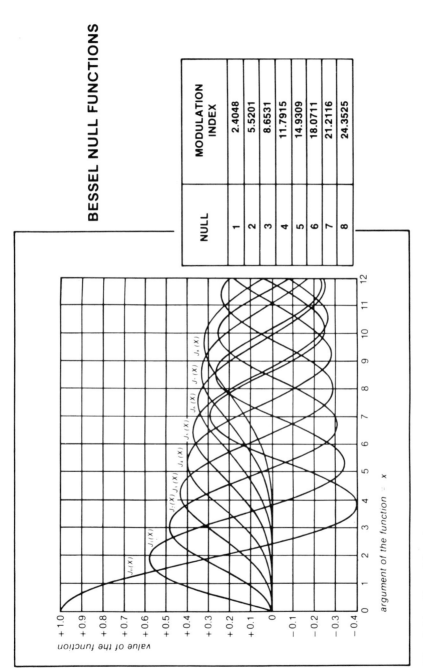

Fig. 2-11. Bessel functions graphed and nulls catalogued in direct values of the modulation index. (Courtesy of Tektronix)

positioned alongside. Such Bessel Functions are single-value integral functions for n = 0, 1, 2 . . . , and are independent solutions of a certain differential equation where, in this instance, the results are of considerably more utility than the equation that precedes them.

What we're looking for are the various null points somewhat between 1 and 8 where the fm carrier amplitude nulls (becomes virtually or precisely zero) and total energy is contained in the sidebands. Depending on the 1st, 2nd, or 3rd, etc., null, the modulation index is then known, and either the frequency or peak deviation may be calculated. For instance, a peak deviation of 5 kHz and a modulation index (1st null) of 2.4048 would produce a frequency of:

$$M = \frac{\text{Peak Carrier Deviation,}}{\text{Highest Modulating Freq.}} \quad \text{so } f = \frac{\text{Peak Deviation}}{\text{Mod. Index}} = \frac{5 \text{ kHz}}{2.4048}$$

or f = 2079.17 Hz. As usual, any two knowns can derive a third related unknown. If you want to find deviation directly, a single audio tone can be used to reveal half the analyzer's filter output and the deviation might then be measured from the viewable display: the number of divisions occupied times the analyzer set bandwidth. To check a transmitter accurately for 100 percent modulation, supply an audio signal to the transceiver, monitored by a frequency counter, and then select a modulation index whose null would be within the instrument's passband. Along with accurate peak deviation, this will determine the 100 percent modulation calibration point (Fig. 2-12).

Indication of the Second Null Using a Spectrum Analyzer

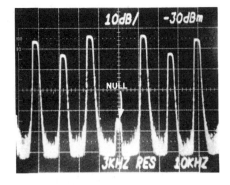

Fig. 2-12. RF test point (transmitter) sees audio generator set for 13.586 kHz input produce 100% modulation at second null. f=75 kHz/5.502. (Courtesy of Tektronix)

OMNI-CHECKS—EVERYWHERE

Power Output Test

This procedure is for any amplifier, and can be at some calibrated tapoff from the transceiver output as well as an audio amplifier. Dial 1 kHz modulating frequency on a signal generator, put it through the equipment under test, and increase its output until visible (at 10 dB/div) on the analyzer's graticule. Continue to increase the generator output until levels of the 2nd and 3rd harmonics increase in amplitude more rapidly than the 1 kHz tone. The point where harmonic distortion increases more rapidly than the primary signal becomes the maximum undistorted power output. Now, switch in 2 dB/div factor and observe the dB down-count from the top of the display. If full screen is 100 watts, then 3 dB down amounts to half power, or 50 watts. Two dB down would be a loss of 0.631 times 100 W, or 63.1 W remaining output.

Another method of measuring undistorted power is to apply a sinewave both to a spectrum analyzer and another oscilloscope. When the sinewave begins to clip, read the dBm power level on the analyzer.

Harmonic Distortion

Harmonic distortion (total) can be found by adding together the amplitude levels of the several harmonics resulting from a 1 kHz tone movement through an amplifier.

Initially, find the power rating, as in the preceding power output paragraph, and then switch back to the 10dB/div and establish both position and amplitude of the 2nd harmonic versus the 1 kHz signal. Whatever the dB amplitude ratio between them is the second order harmonic distortion and can be read as distortion percentage by using Tektronix' Table in Fig. 2-13. Total harmonic distortion sums the RMS of all harmonic levels. If, however, other harmonics are 6 dB or better down from the 2nd harmonic, only the 2nd harmonic need be used in the analysis.

Intermodulation Distortion

This problem is considerably more difficult to find and conquer because it is likely to be intermittent when sought in the real, operational world. In troubleshooting, of course, two rather pure sinewave pairs such as 60Hz and 6 kHz, or 7 kHz, 12 kHz, 14 kHz, etc., are injected into amplifiers to determine how much one frequency interferes with another. The IHFM high fidelity manufacturers use a filament transformer for the 60 Hz source with a mixing network and step attenuator, while the SMPTE motion picture and TV engineers separate the two signals by a voltage ratio of 4:1 so that the 6 kHz tone is down 12 dB from 60 Hz.

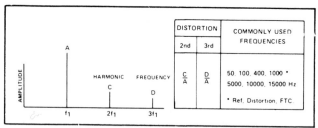

Harmonic distortion method of measurement

RATIO in dB	% of READING	RATIO in dB	% of READING
20, (40;60)	10% (1% .1%)	30 (50,70)	3.16% (.31,.031%)
21	8.9	31	2.87
22	7.94	32	2.51
23	7.08	33	2.24
24	6.31	34	2.00
25	5.62	35	1.78
26	5.01	36	1.59
27	4.47	37	1.41
28	3.98	38	1.26
29	3.55	39	1.12

Chart for Conversion from dB's to Percentage Readings

Fig. 2-13. Chart and tables illustrate both harmonic distortion factors and a handy chart to convert decibles to percentages. (Courtesy of Tektronix)

Amplifiers are adjusted for flat response with volume control maximum. With 6 kHz as a reference, the amplifier is driven to within 14 dB less than its rated power. The sum of both tones will now amount to the amplifier's equivalent rated power. Tektronix then suggests that a 100 Hz/div span and a 10 dB/div be selected at the 6 kHz signal tuned to center graticule. All sidebands around the *base* of the 6 kHz tone are considered modulation components of 60 and 120 Hz and may be added together with certain small corrections (not given). The percentage of intermod distortion may then be read from Fig. 2-13, the conversion chart for percentage readings from dB.

Signal-to-Noise (S/N) Ratio

A much easier measurement than Intermod is the signal-to-noise ratio that's basically the dB difference between a reference level (top of graticule) and the noise floor with the video filter engaged. You're expected, of course, to have a load and series network from any amplifier, with said amplifier's input supplied by the usual 1 kHz sine wave. Increase signal inpuput until the amplifier reaches its rate output with volume control at maximum. The analyzer's bandwidth is then set to 2 kHz/div. and resolution to 3 kHz. At 10 dB setting, the level of the 1 kHz tone is noted and becomes our reference. Remove the 1 kHz tone and short the amplifier input With video filter engaged, the difference between reference level and video filter level is the S/N ratio which you read directly. In audio work, a correction factor between 3 kHz and 15 kHz can be deducted, which amounts to a true reading. For instance, if signal amounts to 10 dB, filter floor 60 dB, and correction 7 dB, the S/N at 15 kHz would amount to −43 dB. If no correction, your answer would be 50 dB. This method is for audio frequencies only. At video or equivalent frequencies you have another very useful method that may or may not involve a weighing filter (EIA specifications) or a highly calibrated signal source and a dual-trace oscilloscope with good sensitivity.

S/N Ratio With Dual Trace Scope and Signal Source

If you're fortunate enough to have a Tektronix 147A or other reliable NTSC generator, then 100 IRE units amounts to 1 volt and there's no interpolation, just straight-ahead measurements. However, many are not blessed with such expensive and highly accurate broadcast equipment, so alternate and somewhat less precise means are often handy, especially for television receivers. But careful measurements taken at the cathode ray terminals should not vary more than 10 percent from any norm. As an example, our calculations in two separate instances using both exact and substitute equipment varied only 4 dB between 42 and 38 dB readings, respectively. As you might expect, the precision equipment results were higher.

In this instance, a VA48A Sencore Video Analyzer (Fig.2-14) having both Bar Sweep and Chroma Sweep modes with staircase for gray level was used. First, the Bar Sweep was engaged for calibration, and then the Chroma Sweep was turned on for final measurements. All instrument RF settings are at Normal.

Initially, a horizontal broadcast sync signal is recorded at the set's video detector, and a black level to sync pulse tip is seen covering 3 divisions at 200 mV/division, establishing a ratio of 40 IRE units for 600 millivolts. With 3 then divided into 40 (13.33), the number of IRE units/div. is set. Three generator staircase steps amount to 5.2 divisions, so 13.33 is multiplied by this figure to produce 69.33. Then, to obtain the number of divisons per measurements, 69.33 is divided into 100, producing a ratio of 1.44, amounting to our 100 IRE units.

The low-capacity oscilloscope probe is now removed from the receiver's

Fig. 2-14. Sencore's versatile VA48 TV video analyzer is also useful for noise measurements in conjunction with an ordinary oscilloscope.

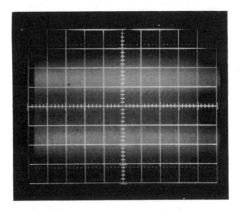

Fig. 2-15. Identical noise into the scope's twin vertical amplifiers must be meshed before final "0" measurement.

video detector and connected to the red or blue cathode terminal on the picture tube.

With the three generator steps measuring 5V/div., this is multiplied by 4.4 div. and then again by 1.44 to reach the final 100 IRE reference of 31.68.

At this point the second probe from your dual-trace or dual-beam oscilloscope is connected to the same point as the first one, with the VA48A set to Chroma Bar sweep. When the resulting noise is meshed at 2V/div and the probes switched from AC to ground, there appears an 0.8V difference. This is now multiplied by 0.5 to obtain an equality with 2 RMS, and 31.68 is then divided by 0.4. Finally, the 20 × log of that number amounts to 37.97 dB, which is a reasonable signal-to-noise ratio at the cathodes of any TV receiver picture tube.

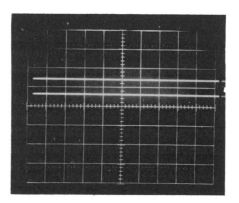

Fig. 2-16. With amplifiers switched to "ground," the noise level amounts to 0.8V between the scope's two amplifiers, and with a few calculations becomes 37.97 dB S/N ratio of this particular TV receiver.

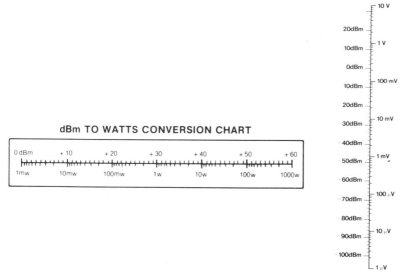

Fig. 2-17. Two dBm (power in dB referenced to 1mW) to microvolts (10^{-6} volts) and watts comparison charts for handy reference between these differing measurements. Often +20 dBm is the maximum input for communications test sets (monitors).

Obviously, a single measurement with the spectrum analyzer would have been easier and probably more accurate, but, nonetheless, you do have an alternate method that might prove very useful under certain conditions where an analyzer isn't available.

SUMMARY

You may be certain there are many, many more spectrum analyzer applications known than are given here, but the foregoing will certainly help toward establishing this highly useful amplitude and frequency measuring instrument as a major aid in both design engineering and troubleshooting any electronics that generate analog signals. It is also anticipated that spectrum analysis will also become an absolute necessity in servicing all cable vision and 2-way radios. There is no other convenient visual means, for instance, of identifying crossmod, intermod, ocscillator, harmonic, and spurious distortions which, in many instances, must meet Federal specs before being approved for communications transmissions. Sixty dB down, in 2-way and CB trade, for instance, is already a way of life, with 75-80 dB down on the way.

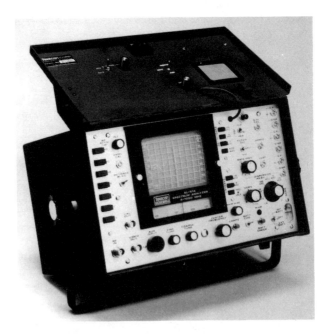

Fig. 2-18. Polarad Electronics announces a new 632B-1 100 KHz-to-2-GHZ spectrum analyzer with on-scale 80 dB dynamic range (100 dB/div scale), digital memory, and 300 Hz initial resolution, plus 100 MHz internal markers and harmonics.

Chapter 3
Logic Analyzers

Very different from spectrum analyzers that produce an amplitude versus frequency X-Y plot, the logic analyzer may be anything from a simple go, no-go probe to a highly complex microprocessor or minicomputer that's needed to "read" the many rapidly developing microprocessor integrated circuit systems presently available on one or more monolithic chips. Tektronix, Hewlett Packard, and Biomation (Gould) are apparent leaders in these efforts and deserve front-runner recognition for the excellent equipment each produces. Unfortunately, logic systems promise to become considerably more involved than they are even now, and test equipment must inevitably follow much the same electronic route. Expect, therefore, many more systems with microprocessor computer motivators, various video readouts, and even entire equipment consoles devoted to analyzing both small- and large-scale logic arrays now in design or early stages of production. In the not too distant future, formerly all-analog systems such as long lines video, telephony, CATV, satellite communications, etc., should all be digitized for power conservation, accuracy, and longer distance transmissions with virtually no detectable errors. Digitizing will also expand the discipline of reasonably secure communications for business, special interests, and government, adding to the popularity of logic processing in any and all disciplines.

In the past, and certainly at present, many of these logic analyzers have been quite expensive, and some rather bulky. Tektronix, however, had just announced a Model 308 (Fig. 3-1) with clock frequency of 20 MHz that combines logic-state, logic-timing, and signature analysis in a single portable package that weighs just 8 lbs. and sells for $3000. Serial logic-state analysis can be synchronous or asynchronous, with either internal or external clock permitting data to enter the analyzer as 5- through 8-bit incoming word lengths in RS-232 format. In parallel timing, the 308 receives 8 channels and has a 252 bit/channel memory and an 8-channel parallel word recognizer already built in, with probe-extending triggering to 24 channels. Stored information shown on the CRT may be of binary, hexadecimal or ASCII formats, with odd, even, or nonparity word recognition.

54 / Logic Analyzers

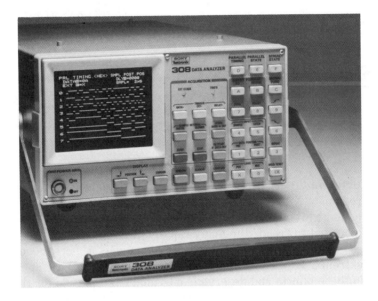

Fig. 3-1. Cost effective, light in weight and portable, the 308 offers logic state, timing, and signature analysis in a brand new instrument. (Courtesy of Tektronix)

In its signature mode, the instrument converts received logic between start and stop gate signals into a 4-digit alphanumeric code. There are also Hold and Repeat modes for storage and input bit stream comparisons, respectively, and the latter mode will also detect any changes. This advantage is present, too, in parallel timing, where storage bits can be compared with incoming information in a "babysitting" arrangement which finds faults and triggers without the analyzer's operator being present.

There's more, of course, but this late instrument's introduction was deliberately included to permit an immediate view of the state-of-the-art now prevalent in logic analyzers. Other very considerable advances by other outstanding manufacturers are sure to follow in the immediate months and years ahead as the electronics of digitizing constantly progresses.

ANALYZERS DEFINED

Hewlett-Packard, among others, chooses to place most analyzers in two main categories—Timing and State analyzers—and this is as good a place as any to begin our specific discussion.

Timing Analyzers

This equipment uses its own internal reference oscillator to sample information, telling whether incoming intelligence is in a high or low state compared with the clock. Timing resolution, then, depends on the sampling clock period and is read out sequentially. Desirable rates of sampling, ordinarily, are some four to six times that of the internal oscillator. According to Hewlett-Packard, "an essential feature of a Timing analyzer is the ability to capture and display glitches . . . that occur within a sample period." This manufacturer then references its Model 1615A, which has a video readout. Such "glitches," of course, are usually narrow spikes of varying amplitude that, if falling at coincident times, can trigger system logic producing false readouts of any description. Errors appear and, if not suppressed, can render the entire logic array incapable of accurate response or reproduction. Simple "latch" circuits that expand these glitches often do *not* "read" ensuing aberrations immediately, especially during level changes (transitions from high to low or vice versa). So more sophisticated electronics are required both to capture glitches and to follow the on-going logic flow. Timing, especially glitch triggering, is highly important to these logic analyzers and should possess most all of the various methods that follow: Asynchronous, Synchronous, External, Delayed, or Armed. In most logic analyzers the systems used are Probe (in contact with the digital system); Memory (of the Analyzer); Trigger (just discussed); and the appropriate CRT or other readout, called a Display. According to Tektronix, all-channel binary information is received simultaneously; such data is stored in memory; stored data is stopped by an input trigger; and then the formatted output is displayed for analysis.

State Analyzers

This clock-strobe system is much like that attributed to the Tektronix description in the preceding paragraph. A trigger or strobe from the tested logic system operates these State analyzers which must read the various word and sequence parameters of the tested system exactly as the information is generated, with all data in a stable state after being contained by an external trigger. Such State analyzers are often directly involved in troubleshooting microprocessors and their associated systems, encoding information into hexadecimal or inverse assembly and other convenient formats. In selective tracing applications, these analyzers concentrate on only special sections of interest in any digital system, with special triggering and certain memory storage. Adjustable triggering then, will permit analysis of *different* bit streams within the digital system, making the selective tracer suitably flexible for many overall applications. High and low counts of successive pulse streams, as well as "nested loops" involving electronics akin to the old mesh currents and branch voltages concepts known originally as *planar networks*, are the techniques extant.

It's interesting to note that *any* oscilloscope can trigger on a special pulse or rise/fall time of some particular waveform, but it will only record an event *after* it has occurred. Conversely, the logic analyzer is able to exhibit events existing before the problem "glitch" or dropout so that conditions preceding the fault may be analyzed as well. Digital data consisting of 4, 8, or 16 channels simultaneously may be viewed by active multichannel probes in the Tektronix equipment with accumulated storage of 4096 bits of memory. This means, according to Tektronix, that you may store 1024 memory bits in 4 channels, 512 bits in 8 channels, or 256 bits in 16 channels. Center trigger and post trigger displays are selectable on a 50/50 and 10/90 percent basis, respectively, and a trigger marker shows what trigger point you're on.

TESTING PROCEDURES

Both Tektronix and HP have interesting diagram approaches that can be used to illustrate quite graphically how such logic test systems are applied, although HP seems oriented more toward design than pure troubleshooting and debugging. Nevertheless, both are excellent examples of very modern techniques that should prove useful under almost any conditions. We'll take Tektronix' offering first.

Selective Triggering

Tektronix says any parallel word having up to 16 bits content may be selected by front panel switches from among Hi, Lo, or X (don't care). Each time the word recognizer detects such stated conditions, it will trigger the logic analyzer and match an incoming parallel word to any other selected. In addition, triggering may also take place on channel 0 or external signals such as errors, flags, sector pulses, enable signals, or other unique single-channel events. In the end, of course, the operator of any logic analyzer must interpret the display and discover for himself any actual errors and their source. In the process, the Tektronix equipments offer "tick" marks to quickly identify such special data bits and also channel position controls that permit bit steam movement between adjacent channels for time comparisions.

In *checking software*, according to this manufacturer, you sample the data flow synchronously, that is, synchronize your analyzer with the system's clock so that all valid data may be inspected, especially during Hi-Lo clock transitions. In examining *hardware systems*, however, Tektronix suggests asynchronous data sampling so that your analyzer operates solely on its own clock, somewhat out of time with that of the system under test. This is how system clock times are found, as well as various other comparisons required for an overall system check. Naturally, data must be sampled frequently with usually very short times between samples for best accuracy. At 100 MHz, the actual timing resolution amounts to 15 nanoseconds; and, asynchronously, that's an extremely rapid rate.

Tektronix' block diagram illustrating external-internal clock gate input to sequential access memory, its external trigger and word recognizer, the multiple channel probe, and the X,Y,Z display generator and triggered input are all shown in Fig. 3-2. In the memory, every incoming bit is stored successively in some special memory location, and when the next bit enters it is stored in the first bit's location while that bit is moved forward in sequence until the memory process is stopped by an input trigger. Pre-center or post-trigger modes then determine the state of the analyzer, and an intensity marker called a *cursor* (from the age of slide rules) will mark special portions of any channel that should be seen.

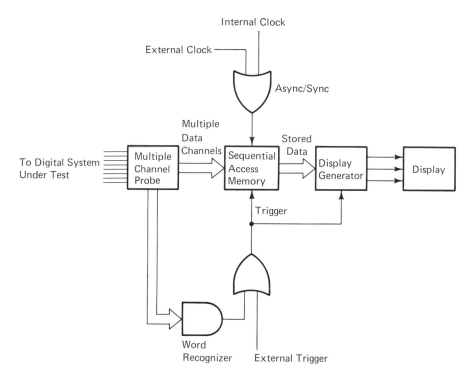

Fig. 3-2. Tektronix block diagram of a logic analyzer showing all prime functions. (Courtesy of Tektronix)

58 / Logic Analyzers

HP's Digital Analysis

The drawing in Fig. 3-3 amply illustrates Hewlett-Packard's approach to real-time digital analysis and shows the "design cycle for a typical digital system," consisting of a microprocessor, minicomputer, or even ROM (read only memory) based controller sans central processor unit (CPU). Software people, of course, are interested in best execution and program areas, while the hardware group wants to know about coupling or "glitch" errors as well as race and trigger conditions. When software

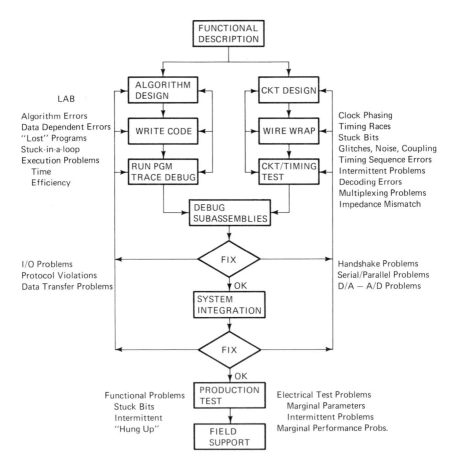

Fig. 3-3. HP's flow chart illustrating microprocessor logic development control in a digital system. (Courtesy of Hewlett-Packard)

and hardware are operated together, then input-output, timing, stacking and related bit streams are all due to be studied for faults that must be corrected. Depot service, HP says, is "not unlike production environment," but there is less emphasis on speed and more on equipment portability so that it can be examined under maximum repair and troubleshooting conditions.

HP then lists three digital system modes of operation in Fig. 3-4. In analyzer language, these three modes are named bit serial, byte serial, and word serial. And only when "word-flow errors" are apparent does the engineer or technician need to worry about voltages that drive or produce these words. Actually, if a single word consists of 8 bits, it can be monitored as 8 serial bits, 8 parallel bits, or 4 and 4 parallel bits.

In using analyzers, one should remember that digital displays (or signals) are not single but consist of many lines because of data, clock, counters, strobes, etc., that already appear on more than one visual line. Some signals occur only once, but many other commands or inputs are repetitious, although not at certain times and during specific periods. So test oscilloscope instruments must have parameters defined according to probes, triggering, and overall display, as well as words in terms of sequence, events, or times, because, unfortunately, there is no positive way to answer precisely what occurs after the initial switch is closed or the first strobe seals memory. An error, unfortunately is found in a mass of data, not among the initial digits of some specific line. The foregoing generally constitutes the *"who, why,* and *where"* of logic analyzers, while this text will, with the help of Tektronix, explain *what* goes on during the process of digital testing in very simple, easy-to-grasp terms.

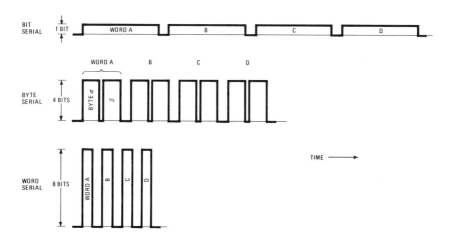

Fig. 3-4. Bit, byte, and word serials are the three digital system types of logic operations. (Courtesy of Hewlett-Packard)

HOW THEY OPERATE

At the input, a voltage comparator (think of an OpAmp with some dc reference level) electronically decides if the incoming voltage is either a high or low, depending on whether it is above or below the assigned threshold. Logic is then discretely sampled so that the various binary states may be stored sequentially in memory in either synchronous or asynchronous modes, depending on the type of problem you're trying to solve (Fig. 3-5). As Tektronix points out, however, the sequential access memory is not just a collection of basic shift registers but, more probably, high-speed random access memories (RAMs) and a counter. Of course, as old information is overwritten, new data automatically stores so that the RAMs always have fresh memory from the 4-, 8-, or 16-channel format being investigated.

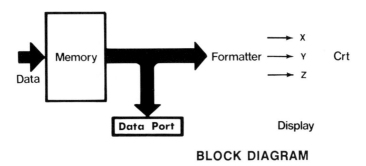

Fig. 3-5. General block diagram of a logic analyzer system from memory to cathode ray tube display. (Courtesy of Tektronix)

Synchronous Sampling

In synchronous sampling, the system's own clock times information of interest to the tester as input data enters. Glitches *between* pulse excursions are not counted, and only 1s and 0s are stored (Fig. 3-6).

Asynchronous Sampling

In asynchrous-sampling (Fig. 3-7), just about the opposite is true. Here a clock in the logic analyzer provides timing that is *not* coincident with system clock (normally faster), and glitches *following* voltage excursions are stored. When the sampling clock rate is actually faster in the system than the test gear, a condition known as "aliasing" appears and is shown as standard transitions and small clock samples like the fourth hi-lo illustration in the figure. There is also a glitch latch which displays any stray voltage between sample clocks as a single clock period pulse during the

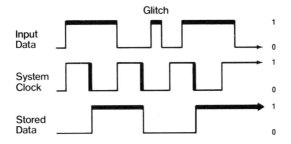

SYNCHRONOUS DATA SAMPLING

Fig. 3-6. Synchronous data sampling. (Courtesy of Tektronix)

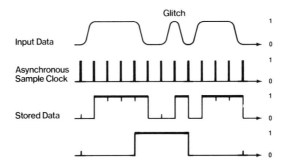

ASYNCHRONOUS DATA SAMPLING

Fig. 3-7. Asynchronous data sampling. (Courtesy of Tektronix)

succeeding clock interval. So when evaluating logic conditions in synchronous sampling, clock rise and fall times, including setup and hold, are the parameters of interest; while in asynchronous sampling, the minimum pulse width tells the false or true story.

Triggering

Word recognition and bit pattern triggering often are highly useful in single-shot and comparison techniques where the logic analyzer is programmed for hi, lo, or X (don't care) conditions to produce a trigger whenever some specifically selected word occurs.

Word Recognition

Synchronously, a clock change becomes the enabling pulse for pattern recognition while any false signals between clock pulses from the equipment under test are not counted. Asynchronously, each time the selected *word* appears, a trigger is produced.

62 / Logic Analyzers

Cathode Ray Tube Displays

There are three of these, and all are illustrated in Figs. 3-8, 3-9, 3-10, courtesy of Tektronix, who supplied all three. The first is a timing diagram (Fig. 3-8) illustrating a display of 33 words, beginning with CH 0 and ending with CH 8 and extending from the 1st to the 512th sample. The little dots are the cursor, or tick, marks, that identify selected bits of information so they can be easily read from the diagram. Because this is a drawing, the sharp positive and negative transitions of each high or low data sample is not shown.

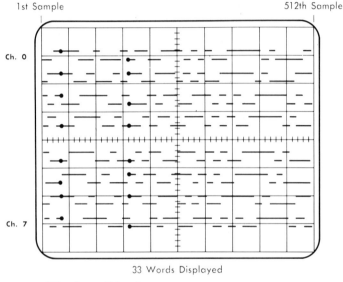

Timing Display

Fig. 3-8. A 512 sample timing display of 33 words. (Courtesy Tektronix)

Binary Display

Once more shown on the cathode ray tube, but this time without graticule lines and presented as an illustrative drawing. As you can see (Fig. 3-9), there are 16 words and 16 channels illustrated, apparently randomly and in hexadecimal format. In octal, of course, the groupings would have been in terms of threes, not fours. Regardless of hex, binary coded decimal (BCD), or octal, each word is in binary 1s and 0s. Such groupings, then, are organized into 3- or 4-bit bytes to suit the user.

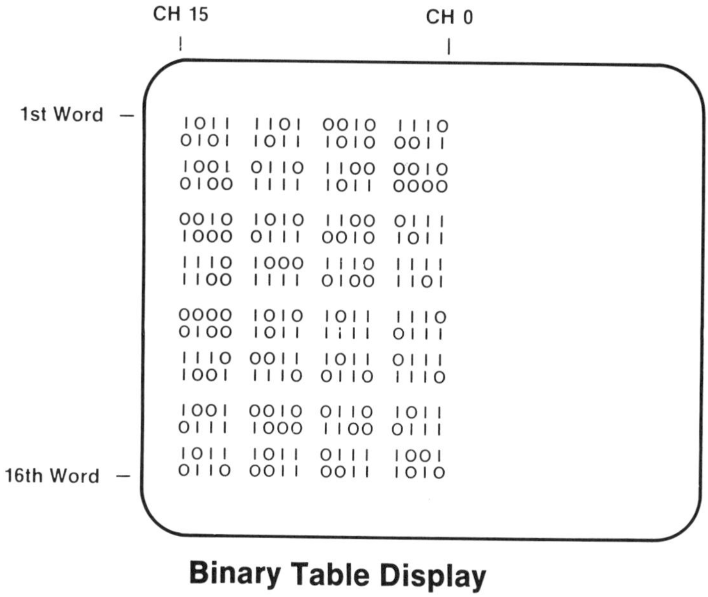

```
               CH 15                    CH 0
                 |                       |
1st Word  —    1011  1101  0010  1110
               0101  1011  1010  0011
               1001  0110  1100  0010
               0100  1111  1011  0000
               0010  1010  1100  0111
               1000  0111  0010  1011
               1110  1000  1110  1111
               1100  1111  0100  1101
               0000  1010  1011  1110
               0100  1011  1111  0111
               1110  0011  1011  0111
               1001  1110  0110  1110
               1001  0010  0110  1011
               0111  1000  1100  0111
               1011  1011  0111  1001
16th Word —    0110  0011  0011  1010
```

Binary Table Display

Fig. 3-9. Sixteen words and 16 channels as a binary table display. (Courtesy of Tektronix)

Map Display

As opposed to binary, this technique of mapping actually amounts to an X-Y display in dot matrix form offering a broadband sweep exhibit of system performance (Fig. 3-10). Much like the junctions (nodes) and branches in network topology, these graphs can actually save troubleshooting effort by quickly indicating a specific data flow in any program, eliminating some of the more time-consuming word and byte logic analysis.

Displayed are logic words of up to 16 bits, with individual dots indicating a single 16-bit word. This dot's position tells the binary magnitude of a specific word, its brightness, and the relative number of times it occurs. Connecting two or more dots is a trace showing both start and stop positions of that particular word located in a special vector.

Hewlett-Packard's 1600A analyzer generates such maps as samples, separating each 16-bit parallel word into two bytes, the most significant of which goes to the vertical deflection plates through a digital-to-analog converter (DAC). The least significant byte goes to the horizontal deflection plates via a second digital-to-analog converter. In the NORMAL mapping operation, however, only the six most significant bits of each

64 / Logic Analyzers

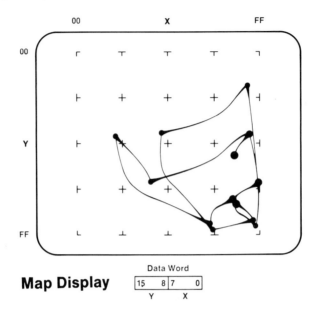

Fig. 3-10. Vector map shown, especially for system quick look checkout. (Courtesy of Tektronix)

byte locate the dot. In an EXPAND mode, only two most significant bits of each byte identify the normal mode quadrant that is to be displayed. Display resolution amounts normally to 4096 states (2^6 by 2^6), and in the expand mode to 65536 states (2^8 by 2^8). In the illustration of Fig. 3-10, logic states shown are in 2^n by 2^n matrix, where n equals one-half the channel capacity.

Mapping changes can be detected easily by trained eyes that will often pinpoint a system breakdown much faster than regular oscilloscope or analyzer analysis, and good system checkout procedures will keep polaroid pictures or drawings of good maps available for immediate troubleshooting comparison. The same suggestions are true for localized investigations, although dot intensities may not always indicate precisely how many times this particular word occurs. Then, of course, tabular display and standard electronic trouble shooting procedures must be used.

When evaluating possible future system difficulties and any additional logic generators you may use, select specific test gear for your foreseeable applications rather than broad, general instrumentation that may not do precisely what is required. Unfortunately for many (and fortunately for some), this is an age of specialization because of the many electronic disciplines, and also-ran test equipment means both lost money and wasted time. Good test gear virtually speaks for itself when produced and marketed by reputable manufacturers. Do read the specs.

GENERAL TROUBLESHOOTING

A classic example of an error apparent is found in a specific study by Tektronix in its 16-channel logic analyzer pamphlet first published in 1975. Here Marshall Borchert uses the LA 501 Logic Analyzer to uncover problems in a defective microprocessor (Fig. 3-11).

This processor, according to Borchert, was in the design stage and would not operate on a very basic program that required simple sequence through the first 16 addresses. Center screen triggering was chosen to exhibit events prior to the trigger point, and 8 inputs were entered. At 200 nanosecond intervals the LA 501 synchronous clock was set, since the shortest pulses in the microprocessor lasted for 1 microsecond. Now, 512 samples can be viewed at the 8 inputs, and as the illustration shows, certain bits are showing up longer at certain intervals where they should not, according to previously drawn logic diagrams. The circled portions of the four illustrations indicate that shortly after the end of μP housekeeping and start of the actual program, an elongated bit appears on CH 3. This is expanded first horizontally then vertically, and it shows trace expansion also at channel 6. Example D is an identical display to that of C, except that a new microprocessor has been installed and all sequences are now correct as designed. Note that channel 3 has been moved down so that its coincidence with channel 6 is immediately apparent. Also observe that the CH 3 input pulses are not a pair of highs but a single high and the inverse of CH 1 and CH 2, remaining initially on, then in an *off* state for as far as we can see in D. (Presumably this is positive logic.) Unfortunately an ordinary oscilloscope with but two normal channels and conventional triggering could not and would not illustrate this significant train of logic pulse events. An external oscilloscope electronic switch wouldn't do the trick either because of the 200 nsec time base element and also the inability to stretch such pulses into an expanded display. So for starters, this is a pretty good example of the inescapable usefulness of a good logic analyzer.

A 7D01 Analyzer Plug In

The 7D01 (Fig. 3-12, 3-13) converts any 7000 series Tektronix oscilloscope into a formidable and formattable logic analyzer. Sixteen 3-position switches, in addition to others for sync, external qualifier, probe qualifier, polarity, and memory provide flexibility for this dual compartment unit, which also has sampling interval setting of from 5 milliseconds to 10 nanoseconds. The two probes are active multichannel signal contacts,

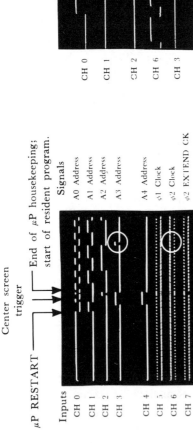

Figure A. Unexpanded replicas of 100 µs segment of eight binary signals, recorded simultaneously on a one-shot basis.

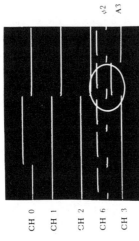

Figure B. Horizontal expansion of same signals as in Figure A. positioned to keep regions of interest on screen.

Figure C. Vertical expansion with φ2 positioned directly above A3. The A3 pulse should not end when it does.

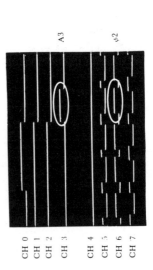

Figure D. Same as Figure C. except the defective microprocessor IC is replaced by a good one.

Fig. 3-11. Actual problems in a microprocessor-controlled logic system. (Courtesy of Tektronix)

General Troubleshooting / 67

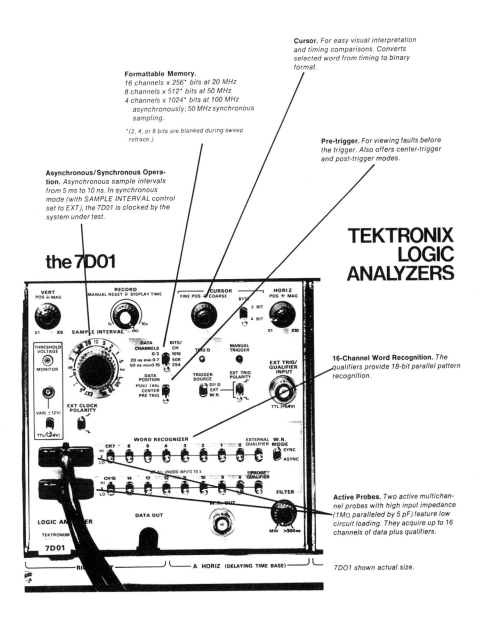

Fig. 3-12. Tektronix 7D01 do-all logic analyzer plug-in for all 7000 series oscilloscopes. (Courtesy of Tektronix)

68 / Logic Analyzers

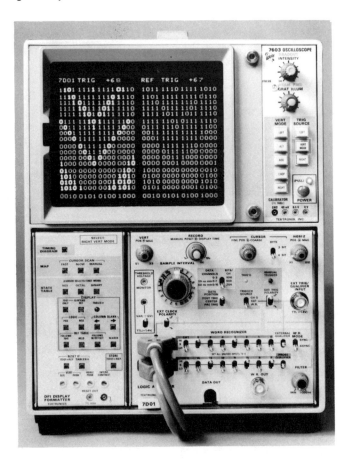

Fig. 3-13. The 7D01 logic analyzer teamed with the DF1 display formatter provides a powerful team in logic analysis. (Courtesy of Tektronix)

each sampling eight channels for a total of 16. Tektronix also has an LA 501 logic analyzer package (Fig. 3-14) that works in any TM 500 mainframe to turn companion oscilloscopes into highly effective analyzers. Generally, like the 7D01, the LA 501 will monitor 16 channels, has active probes, and a distinctive channel positioning control, plus qualifier and clock.

General Troubleshooting / 69

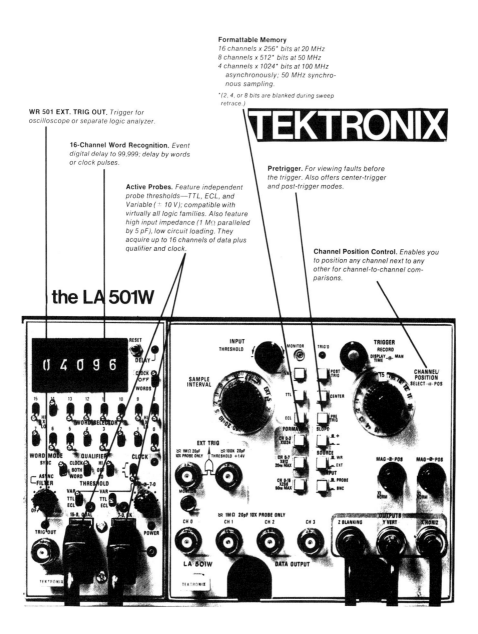

Fig. 3-14. The companion LA 501W logic analyzer, used mainly with T500 oscillocopes. (Courtesy of Tektronix)

Chapter 4

Storage/Sampling Oscilloscopes and Time Domain Reflectometry

Storage and sampling oscilloscopes are by no means equivalencies, but they do complement one another in high and very high frequency ranges, even though sampling is considered somewhat restricted rather than a complete intelligence measurement. Time-domain reflectometry, however, is now an electronic adjunct to sampling instruments, by at least one manufacturer, and requires a certain amount of discussion, especially where various cable transmission systems are involved. Consequently, this chapter will offer somewhat concise but (we hope) adequate explanations of all three types of scopes, including specific details of their applications, especially the very fast rise times of the sampling/TDR vertical subsystems that, as usual, are priced accordingly.

Each of these oscilloscope test systems, of course, has its specific uses and must be operated in the environment for which it has been developed. Storage is needed for those displays that require study time and analysis. Sampling becomes handy in situations where intelligence is repetitive and arrives very rapidly, and TDR (time domain reflectometry) simply enhances the art of cable system investigation and evaluation. Although you may not intend to use any of these three electronic analyzers in the foreseeable future, an understanding of their virtues and deficiencies could have an important bearing on some unforseen design or troubleshooting consideration. It is always best to be ready for most eventualities whenever possible. We hope the ensuing discussions aid substantially in such preparations.

Specific models and name brands of these oscilloscopes are identified, along with their individual specifications, as a guide of what to look for in evaluating current instruments and not arbitrarily as an advertisement for the manufacturer. Frankly, we know of no other straightforward way to get the message across other than to illustrate what you should expect when selecting (and using) these high-technology instruments.

STORAGE OSCILLOSCOPES

A storage oscilloscope (Fig. 4-1) does exactly what its name implies: it receives information from some analog, digital, industrial, laboratory, or consumer source and stores such impulses for momentary or protracted examination in original form on the face of the cathode ray tube, and can even display intelligence *after* an input is totally disconnected. Persistence storage duration is also often variable, permitting new information

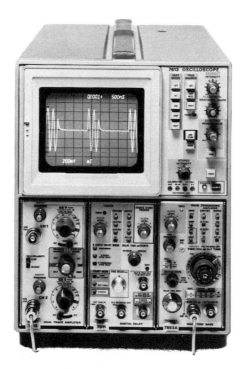

Variable Persistence Storage
4.5 cm/μs Stored Writing Speed
Dc to 100 MHz Bandwidth
Burn Resistant Crt

Fig. 4-1. Mainframe and triple plug-ins make this Tektronix storage oscilloscope a very versatile instrument indeed. (Courtesy of Tektronix)

72 / Storage/Sampling Oscilloscopes and Time Domain Reflectometry

to appear and be examined at the operator's convenience. Automatic trace erasure can also be made available, depending on what you want in the instrument and how much you're willing to spend. Tektronix has a number of storage instruments, and Hewlett-Packard also has several portable 100 MHz variable persistence versions with auto store, auto erase, and 100 cm/usec writing speeds in its 1740 series. Gould, Inc., markets a 10 MHz OS4000 digital storage oscilloscope (Fig. 4-2) with random access memory and can store signals up to 450 kHz. Such information goes through an A/D converter operating on time base setting, and the digital output of this A/D passes into an 8-bit-times-1024-word memory. When such digital intelligence is read out, it moves through a D/A converter and on to the cathode ray tube where it is continuously cycled, simulating a stored trace on the scope's CRT.

In the NORMAL mode, the OS4000 behaves like an ordinary 10 MHz dual-trace oscilloscope having sweep speeds from 0.5 sec/cm to 0.1 usec/cm. In the REFRESHED position, the CRT display reads out stored intelligence, which also reflects any changes in the signal input. In ROLL, the displayed waveform's right edge is in real time, and the trace moves across the CRT according to time base setting. In STORE, a single-shot cycle holds the succeeding triggered sweep, and this trace will then be retained until a release or store button is pressed. In the STORED-TRIGGERED-POINT-POSITION the CRT displays pre-trigger intelligence and all post-trigger information thereafter, with four

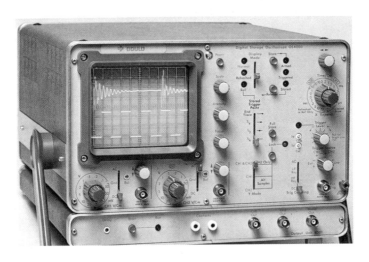

Fig. 4-2. Gould's OS4000 has digital storage with random access memory, plus 10 MHz bandpass and optional printout. (Courtesy of Gould)

trigger points available. Stored trigger points are identified by a bright dot. Gould also offers an OS4100 dual-channel oscilloscope having T-Y and X-Y storage with the same 8-bit and 1024-word RAM plus an option for hard copy recording. Bandwidth is a nominal 500 kHz, and vertical deflection factors range from 100 μV/cm to 5V/cm, with sweep speeds between 100 usec/cm to 50 sec/cm.

More conventional storage oscilloscopes where the CRT is really the storage unit are, however, the most popular equipment around the U.S. and most of the remainder of this chapter will be devoted to their descriptions and uses.

REASONS FOR STORAGE

Surprisingly, there are many! And we'll list a few at random as the applications occur: monitoring for the inevitable transient; examining noise in detail; checking linear detection; marking signal changes during troubleshooting or circuit adjustments; assessing rise and fall times leisurely; viewing a slow signal pattern; stopping any signal for photography and permanent record; looking at signal/noise ratios (done best with a spectrum analyzer having storage); tracking one-shots; and comparing any signal with an already established standard. Obviously, if you think about it, there are many more uses for storage than have been considered in the foregoing lines and, perhaps, we'll find some additions when photographing "live" situations later in the chapter. Whatever the final results, be assured there wouldn't be so many storage oscilloscopes on today's market unless they had already proved highly useful.

STORAGE FEATURES

Bistable Storage

Bistable storage (Fig. 4-3) or split screen storage, permits information to be retained independently on the upper and/or lower halves of the surface of the cathode ray tube and is especially useful where reference and active signals may be compared as separate but related entities. Standard phosphor target bistable displays offer slow writing speeds and rather poor contrast between the trace being viewed and CRT background, but the method is both attractive and relatively low cost.

Bright Bistable

There is also bright bistable available which allows bright, rather long time-duration displays at reduced contrast, made possible by inserting a (screen) mesh between the CRT's gun and its faceplate phosphor.

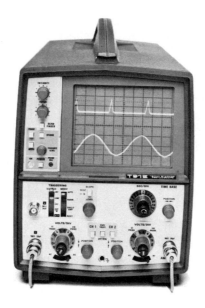

Fig. 4-3. Tektronix's T912 low cost bistable storage and real time 10 MHz offering. (Courtesy of Tektronix)

Variable Persistence

Variable persistence storage is another method of allowing sharp traces to be viewed where either a camera or repetition rate flicker would be apparent under ordinary viewing conditions. Times between virtually instantaneous signal disappearance and storage retention of more than 15 seconds are offered, plus a 100/cm/usec writing rate that is said to be comparable to 3000-speed film. Further, Hewlett-Packard claims that the 1800 cm/usec writing rate of its 100 MHz model 1744A is faster than ASA 10,000 film "without special techniques such as post fogging." In using storage scopes, you must remember that high writing speeds and slow scan often result in some CRT blooming, while slow writing and fast scan rates usually produce a rather pale, dim display.

Variable persistence scopes have a single mode of storage operation but permit adjustment of image viewing times (Fig. 4-4). One display may also be stored for further viewing, if needed, while another waveform is taking shape on the cathode ray tube. In fast, repetitive images, for instance, storage persistence may be adjusted so the CRT will accept and display many traces before the initial waveform fades from sight. Such scopes are particularly useful in evaluating slowly changing electrical pulses or sinewaves that would be only moving dots on real time oscilloscopes. When variable persistence on these instruments is not used, high-contrast stored images are possible for long periods

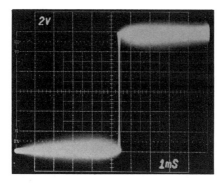

Fig. 4-4. An example of variable persistance for bright, high contrast displays. (Courtesy of Tektronix)

without the usual fading induced by variable persistence. Erasure time for a variable persistence scope such as Tek's 7613 is 0.5 second or less. Auto focus takes care of changes in beam intensity after the main focus control is manually set. Bandwidth for this unit and its 7A18, 7D11, and 7B53A plug-ins is dc to 100 MHz.

Fast Mesh Transfer

For fast mesh transfer instruments are personified by Tektronix 1000 cm/usec writing speed 7633 series of storage scopes that also have a dc to 100 MHz bandpass. This oscilloscope uses high-speed target and mesh-to-transfer techniques allowing longer waveform retention than other storage types, and its waveforms may be seen for hours without loss from fading. Variable persistence is not less than 0.5 minute. Pulses with as little as 10 nanoseconds rise time can also be seen, and these pulses may be fixed on the CRT for weeks if required. A photo of the 7633 is illustrated in Fig. 4-5. Both the 7633 and 7613 are dual trace, and both have the same 7D11 digital delay and 7B53A dual time base; but the 7A18 and 7A26 dual inputs are different. Although both feature constant bandwidth for all deflection factors, the 7A26 has an extra switch and circuitry to limit bandwidth to 20 MHz for low-frequency inputs.

Fast Multimode

Fast multimode storage, according to Tektronix, includes all previously named retention features, plus fast bistable and fast variable persistence where image electrons initially strike a fast mesh before reaching a long retention mesh and the cathode ray tube. Tektronix points out that variable writing speeds of "5.4 cm/usec are increased to 2500 cm/usec by selecting fast variable persistence." This is fast enough to record a "single event" within a bandwidth of 400 MHz or 900 picoseconds rise time, and

76 / Storage/Sampling Oscilloscopes and Time Domain Reflectometry

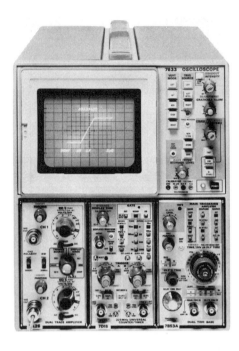

Fig. 4-5. A much faster multimode 7633 mainframe storage oscilloscope having 100 MHz bandwidth. (Courtesy of Tektronix)

that should be sufficient for virtually all applications, even among most superspeed logics. The 7834 oscilloscope that has this remarkable bandpass and writing speed also offers fast multimode storage as well as auto-erase for more frequent image display, a "save" circuit for 30 times additional length viewing, gated readout to stop blooming, and special multitrace delay.

WHAT STORAGE SCOPES DO AND WHY

Glitch hunting with an ordinary oscilloscope or logic analyzer can often seem painful since glitches are not always repetitive and may become random, depending on what's occurring in the investigated system's circuitry. Logic oscilloscopes—if your problem is digital—are a great advance toward positive solutions, but glitches in video analog circuitry are extremely difficult, especially when they fall within the picture and not the sync portion. On the other hand, there are times in digital electronics when these spurious troublemakers don't respond readily to normal scope investigations either. So your only possible recourse has to be among the storage oscilloscope fraternity, whether you're enthusiastic or not. Once you take the plunge, however, the "water will be fine," and most such problems should be readily found. For example, by setting up a 912 Tektronix—which is a comparatively low cost instrument, among the many storage scopes available—we found two positive-going transients riding on a negative-going pulse, upsetting the equipment at relatively regular intervals. Now all you have to do is to expand one of these spurious voltages, measure its width and, perhaps, rise and fall times, and you'll know what time constant or clipping circuit is needed to take it out. The pulses referred to, of course, are between the 5th and 6th and 9th and 10th vertical graticule divisions (Fig. 4-6). Oscilloscope adjustment is done in this instance on a 1-shot basis which can be repeated several times during each exposure until whatever you're looking for appears. In this instance, naturally, positive spikes of voltage.

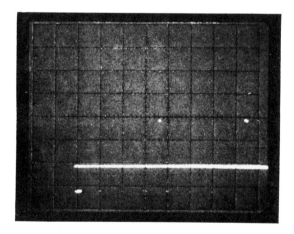

Fig. 4-6. Spurs of unwanted positive voltages appear following a sync pulse.

Let's take another example that's somewhat similar (Fig. 4-7), except that this time spikes of unwanted voltage have been removed and only the pulse is of interest. Done in two successive storage attempts, the pulse appears in good detail at 2 milliseconds/division, making this negative going waveform excursion just about 1 millisecond wide. Of course, at this width, an ordinary oscilloscope with a fast time base can stretch its duration sufficiently to measure rise and fall times with excellent accuracy. Observe that the trace is relatively bright and the background is sufficiently light to see all graticule lines plainly, even though this particular scope has no graticule lighting.

One of the effects you'll notice among the lower frequency storage scopes is the lack of full trace retention once frequencies become just a little faster than storage can accept. In the first photo of this 2-shot series

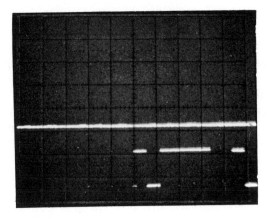

Fig. 4-7. Now, only the pulse is of interest.

(Fig.4-8), you'll see a fairly clean group of sequential burst pulses that are shown in standard real time. But in the second picture, using storage (Fig. 4-9), the outlines of these bursts are shown but any real intensity is saved for those parts of the waveform whose duration is longer than the simply up/down portion of the trace. Consequently, the faster the frequency (and therefore time base setting) the fewer such burst voltages can be stored. So in choosing an instrument that will have maximum utility, be sure its writing speed and storage capability is positively sufficient. One or two trials with the equipment you'll be working with should tell the tale. If a personal opinion can help, don't ever buy test equipment without initially trying it first. It may not really be "the thing to fill the need." Naturally there are certain "tricks" available with the intensity and enhancement controls, but in the end, the storage oscilloscope you want should do the entire job needed without gimmicks or forcing measures, since these indicate the probability of no reserve for those critical voltages that must occasionally be examined. In storage scopes, it's always better to overbuy than underbuy, since marginal equipment is often worse than useless.

As with digital logic, the use of storage scopes will probably grow at a

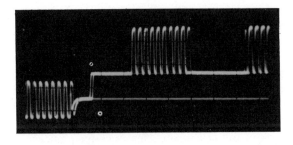

Fig. 4-8. Sequential burst pulses at two levels.

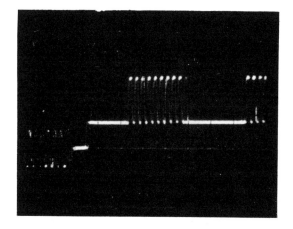

Fig. 4-9. Burst pulses fade with faster time base settings.

considerable rate when such broadcast features as Teletext and Viewdata are approved by the FCC and put on the air and cable. A great deal of work continues in this country and Europe on both such types of video words and graphics, and it shouldn't be too long before they are part of the U.S. air cable transmission system. Storage applications, of course, will arise when portions of the graphics are missing, and you'll have to find the immediate and specific cause. Video information above the traditional 4 MHz baseband will require much faster equipment to resolve processing deficiencies. Once again, "try before you buy." A little intelligent effort goes a long way in appropriate instrument selection.

Broad trace outlines with lower frequency oscilloscopes, however, are very easy to handle, and stepped functions such as those illustrated (Fig. 4-10) do, indeed, require little more than a fair instrument to display their

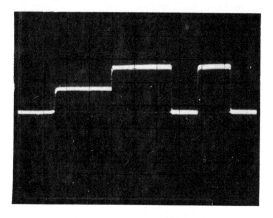

Fig. 4-10. Stepped functions.

80 / Storage/Sampling Oscilloscopes and Time Domain Reflectometry

outlines. With both short-term and long-term information, though, you will need a rather fast storage instrument that can show both. This we'll investigate in the next series of waveforms using a Tektronix 7613 mainframe, a 7A26 vertical amplifier, and 7B53A time base. Here, the mainframe supplies storage, while the two plug-ins do the rest for at least 100 MHz capability.

FAST SCOPES AND STORAGE

Here, we're working with several thousand dollars worth of precision equipment that is built exclusively for laboratory use and not intended for ordinary portable service functions. Highly effective pushbuttons are available for STORE, NON-STORE, ERASE, SAVE, PERSISTENCE, SAVE TIME, STORED INTENSITY, etc. The vertical deflection amplifier accepts inputs up to 200 MHz, while the time base triggers and displays voltages down to 50 nanoseconds—a formidable combination, especially when storage is included also. A 1-shot is available here, too, but it's better to use the more convenient STORE mode most of the time, especially with rapid, repetitive signals, although continuous image buildup can soon distort the trace since the scope continues to store throughout the signal scanning process.

To give you an idea of the power of a 100 MHz instrument, look at the nonstorage trace with the time base set at 5 usec/div. (Fig. 4-11). Every one of the 9 sine wave oscillations following the ON (down) portion of the trade is plainly visible, along with dB and linear markings situated on either side of the graticule. Then, let's use the storage mode and expand the time base setting even further so that the repetition rate of this signal is measured at 250 nanoseconds/division (Fig. 4-12). Note that full detail

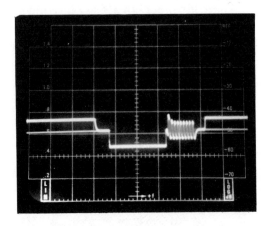

Fig. 4-11. Nonstorage, with time base set at 5 usec/div.

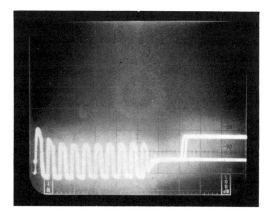

Fig. 4-12. Storage mode at 250 nsec/div. Trace on high frequency scope is very bright and relatively sharp with full detail.

remains even though the trace is somewhat thickened due to repetitive storage scans; but also observe that each portion of the trace is distinct, and there are no distortions except for the second overlap of a stepped section to the right.

With such demonstrations it is not difficult to understand the very considerable difference between fast and slow storage oscilloscopes, and why one must pay a great deal more for better vertical and time base response. Typical rise time for the 7613 with a 7A26 vertical plug-in would be in the very low nanoseconds (2 to 4), equivalent to specifications of the plug-in itself. Compare this with the 35 nanoseconds (tr) of lower storage or standard servicing scopes in the 10 to 12 MHz bandpass categories. As you can see, bandwidth isn't quite proportional to cost, but there is a relationship, that depends, of course, on the going market value of currency and how it affects instrument prices.

While we won't address the bistable split-screen at length, stored and real-time information can be viewed together—on the bottom half, provided that they are of approximately the same frequencies. Mentally put the two prior illustrations together and you can visualize the results. Theoretically, a dual-beam oscilloscope with entirely separate time bases could look at different frequency traces, but the cost would probably be prohibitive, especially where bandpasses are in excess of 75 or 100 MHz. Even at 100 MHz, the better storage oscilloscopes offer but 4.5 centimeters per microsecond, and stored traces are only visible in many for a maximum of 60 minutes. Selection of such instrumentation is a little tricky, and complete specifications should be studied carefully before acquisition, even though most major manufacturers' specifications can be reliably believed. Tektronix, for instance, very often underrates its in-

82 / Storage/Sampling Oscilloscopes and Time Domain Reflectometry

struments, so their delivered performance is an advantageous surprise (fact, not fiction).

SAMPLING OSCILLOSCOPES

Sampling scopes (Fig. 4-13) can be thought of as display instruments which are very useful in high and ultra-high frequency applications, and help to fill any void left by storage and ordinary laboratory-type oscilloscopes. Used as instruments to inspect and analyze waveforms having periodic or varying motions, electrical samples are gathered at very rapid intervals (Fig. 4-14) when the scope's sampling circuit is turned on, waveform amplitudes are measured, and samples are expanded and then amplified at lower frequencies. Since the sample "dots" rapidly appear on the scope's cathode ray tube, the resulting waveform is fairly smooth in appearance and is little more than a lower frequency representation of the original intelligence in somewhat less detail.

Fig. 4-13. Examples of sampling and sweep units used in Tektronix 7000 series oscilloscope mainframes.

Fig. 4-14. Sampling a tuned circuit.

In wideband scopes, the actual instrument bandwidth is determined by the sampling head which accepts signals only in the low volts or millivolt ranges, although 10X and 100X attenuator heads are available to permit sampling instruments to handle larger inputs linearly. Sampling frequencies up to and including 18 gigahertz (GHz) have now become common with advanced technology by several manufacturers. Tunnel and other fast-switching "tender" diodes in the sampling heads account for such low-voltage inputs. Rise times, of course, are in the low picoseconds (less than 350 ps being typical for a 1 GHz sampling head), while 18 GHz are normally specified at less than 28 ps. Plug-in heads for both vertical deflection and horizontal time base seem fairly universal among oscilloscope makers of higher frequency instruments.

Time bases for these scopes are very different also, compared with servicing or laboratory scopes. You may expect characteristics such as 5 usec/div. to 1 nsec/div. in 12 calibrated postions and 10 nsec/div. to 50 usec/div. for the 18 and 1 GHz specifications, respectively, with 3% accuracy typical. All waveforms sampled, however, must be repetitive.

Considering available equipment, it's noted that Tektronix has an interesting 7S12 general-purpose sampling head and time domain reflectometer (TDR) combination in a twin-size plug-in package with system rise time of 35 picoseconds and TDR specs of 1.5 nanoseconds or less for the displayed reflection from any short circuit cable or line. Full-scale TDR ranges up to 4900 feet, with a separate movable tape calibrated for both time and distance, give the TDR portion an accuracy of within 1 percent. If you're working with a combination of very fast information interconnected by cable, this particular plug-in might be well worth investigating because of available options.

TIME DOMAIN REFLECTOMETRY

Since the TDR term (Fig. 4-16) has already been mentioned, a short description of this very useful method of locating cable opens, shorts, coupling, and impedance problems should be helpful, even though you may have little use for it other than general information under ordinary conditions. When, however, the time comes in communication, shipboard installations, master antenna systems, CATV, local and long distance network connections, and multiunit placements in offices and homes of today, and tomorrow, there is no substitute for time domain reflectometry. These instruments are display-type oscilloscopes that generate impulses, voltage steps, or RF bursts with extremely fast rise and fall times. The idea is to transmit a particular signal and then measure the time/foot it takes to return and be shown as a reflected image on the face of the cathode ray tube. In solid polyethylene dielectric with center conductor, pulse propogation rates are approximately 1.5 nanoseconds/foot;

Fig. 4-15. Tektronix 7S12—a 45-picosecond TDR or 30-picosecond sampling head combination. (Courtesy of Tektronix)

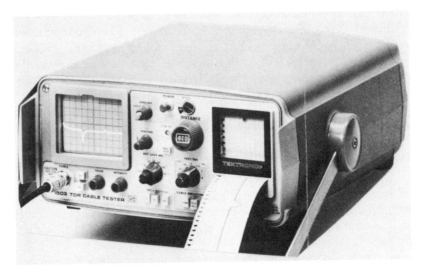

Fig. 4-16. Tek's 1503 cable test oscilloscope with printout. (Courtesy of Tektronix)

therefore, the outgoing pulse and its return consume some 3 nanoseconds. By basic arithmetic, this same pulse can travel 0.1 foot and return in 300 picoseconds. Consequently, problems approximating 0.5 inch may be located within short runs of cable.

Like square waves, step voltages can examine the greatest bandwidths, but their desirable characteristics are attenuated over longer distances, and unless there is high power behind them, their reduced reflections tend to become lost in system line noise. Therefore, TDR equipments used for long stretches of cable often operate with peaks of voltage having expandable impulse widths varying from 10 nanoseconds to 1 microsecond, the narrower voltages being attenuated more than wider ones. Pulse widths, however, must be only a fraction of the distance propogation times. Loss lines, too, require wider pulse widths so impulses can be seen when reflected.

As Fig. 4-17 shows, stepped pulses are usually generated by 50-ohm sources and reflections read on a vertical scale calibrated in terms of ρ (the Greek letter Rho). The load (R_L), between 0% and -100%, and from

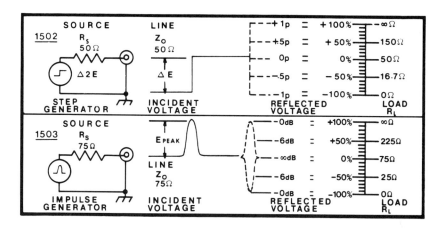

Fig. 4-17. TDR transmission examples at both 50 and 75 ohms. Note reflected signals and their tell-tale loads. (Courtesy of Tektronix)

50 ohms to infinity, by 0% to +100% then is represented by 50 ohms to 0 ohm. Delta changes in the magnitude of this step and distortions of the configurations illustrate system cable irregularities. On the other hand, transmission lines such as 75-ohm video cable accept half-sinewave incident (originating) voltage, and return reflected voltages that may extend from infinity to zero because of the load, which means +100% to −100% in terms of percentages. Consequently, as the dotted reflected voltage indicates, this waveform can be either positive or negative, depending on load (RL) impedances greater or lesser than 75 ohms. In either instance, when loads are reflected back at their *characteristic* impedances, the return voltage is 0% and becomes little more than a straight line. For conditions of opens, shorts, frayed or pinched cable, (Fig. 4-18 through 4-21), the return waveforms will be anything but straight lines and problem eliminations specifically indicated by their electrical shapes. Line terminations, couplings and impedances in time domain reflectometry, of

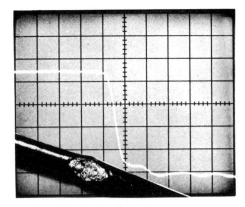

Fig. 4-18. Cable with braid shorted to center conductor. Note downward pulse slant with no rises. (Courtesy of Tektronix)

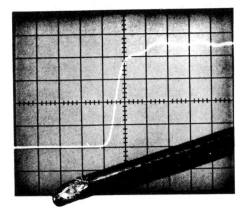

Fig. 4-19. An open cable causes pulse waveform to rise and remain in its high and open position. (Courtesy of Tektronix)

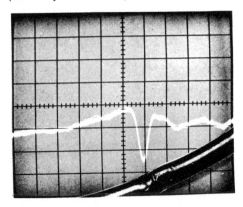

Fig. 4-20. A crimped cable results in an impedance change but nothing catastrophic. (Courtesy of Tektronix)

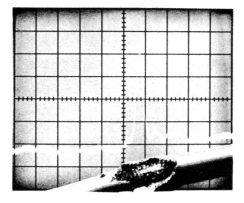

Fig. 4-21. A frayed cable, similarly, isn't catastrophic either, but the impedance change here is twice that of the crimp in Fig. 4-19. (Courtesy of Tektronix)

course, are everything, since discontinuities are the reflected images you're looking for. All reflections, unfortunately, will not always appear since narrowband reflections sometimes are missed due to TDR low-signal energy output. Fortunately, however, this applies more to filters and waveguides, and most cable systems can be completely checked.

Such reflections are exhibited on TDR measuring instruments as time/distance rather than frequency, although rather expensive FDR (frequency domain reflectometry) is available for a price and operates by sweeping some signal generated at a certain frequency and in specific passbands. It will test waveguides and narrow band systems. There is also time domain metrology (TDM) which is supposed to produce all the intelligence available from both TDR and FDR by using a minicomputer for signal processing. But this won't test waveguides and costs a bundle. The most expensive of the time testers—three dimensional reflectometry (3-DR)—is very expensive, quite heavy, won't operate on batteries, but will handle both waveguides and cable, along with locating multiple faults.

Conventional TDR releases either a very fast risetime pulse, a half-sinewave, or broadband burst of microwave energy at about 3 GHz, which will test system cables over a number of frequencies. Reflected images appear at the center on the TDR's graticule and are read in terms of distance—whether in feet or inches. Even multiple faults can be seen and their separate characteristics fully identified, as these regular, repetitious pulses continue as long as appropriate TDR controls are engaged.

For permanent records, a plug-in strip chart recorder is available (with Tektronix), and appropriate X-Y recorders may also be connected, with recording costs less now than "instantaneous" photographs. Scan durations are automatically adjusted in this equipment to correspond to paper speeds in its optionally supplied recorder. The instrument is also able to flood the cathode ray tube during retrace instead of blanking the CRT for convenient image photography without graticule lighting. The process of gating a high-frequency oscillator to control retrace speed is both desirable and necessary, since battery operated TDRs don't have extra current for additional features other than to operate the TDR itself.

Setup and operation of time domain reflectometry equipment is very straightforward, requiring no more effort than attaching cables and pushing selected buttons for various scales and readings. The few steps needed are covered in specific/instrument operating manuals.

CONCLUSIONS

True, this chapter describes the operation and application of highly specialized equipment that ordinary engineers and technicians may use only occasionally or never; however, the knowledge that such instruments are available and what one does with them when needed can be of more than considerable help when those crucial occasions arise. CATV, MATV, cable installers and microwave people, especially, can derive a great deal from the applications illustrated, even though most are general rather than fine-tuned for a single discipline. On the other hand, those working with such types of equipment every day have probably learned things about their own individual electronic pursuits that are often overlooked in universally applied examples. But what you have here should become the basis for a thorough understanding of their usage. As time progresses, uses for such equipment will readily expand due to the multitude of professional, semiprivate, and private installations resulting from transporatation constraints and energy shortages. Fiber optics, high-speed digitizing, a great increase in all sorts of cable runs, and the advanced use of video terminals of all descriptions for computer readouts and all other types of visible images will assuredly require every maintenance and engineering aid available. Storage/sampling and TDR sciences (and arts) are decidedly among those currently ready to do the necessary check-out procedures wherever required. We hope these concise explanations and waveform examples have helped.

Chapter 5
Investigating Video Terminals and Cassettes,
Including the art of using Vectorscopes

Normally it's hardly customary in a textbook to include an entire chapter on specific waveform interpretations. But the overwhelming prominence in sales and service of color video processors and terminals throughout the United States should warrant more than a cursory discussion, especially considering the enormous amount of new and advanced equipment now reaching the market that makes everything possible. The interesting part of all this is that procedures applicable to baseband monitors are generally appropriate for standard receivers with the exception of course, of inevitable phase shifting that always occurs with the addition of extra signal processing, tuned circuits, and the use of direct (baseband) excitation rather than RF.

So with baseband-RF disparities taken into account, following the various RF-IF detection systems, let us begin to consider what actually occurs in some of these video-chroma processors and how specific test equipment can handle both luminance and chroma problems that inevitably arise with electronic malfunctions.

Basically, we'll use a late-model home television receiver as our initial demonstrator, which is generously supplied with integrated circuits and also enough discrete devices to furnish an adequate IC-transistor mix. Then, after video-chroma demodulation and recombination, the sweep, signal amplification, high voltage, etc., will be much the same as any video monitor or video terminal in existence. Individual or proprietary circuits, of course, will not be discussed, since there are far too many of these already on the market, and engineers and technicians must do their own analysis on them.

All fundamentals, however, are covered so that surprises should not be a severe problem. The signal generating test equipment used will consist of three entirely different pieces: the VA48 video analyzer manufactured by Sencore, Inc., a 1248A color bar "sidelock" gaited rainbow generator; and an NTSC Mod. 1250 color and dot-hatch generator, both manufactured by B & K-Precision.

Later we will investigate luminance circuits as well as the effects of combined chroma and luminance so that you will have a well-rounded presentation of most facets of luminance-chroma analysis and servicing. But first, let's look at NTSC signals through a receiver-monitor, since these are specifically designed for chroma processing and demodulation.

CHROMA AND THE VECTORSCOPE

Now the secret's out. What you will initially see are National Television System Committee System signals put to and through a receiver and the results "read" by a vectorscope connected to the cathodes of the set's picture tube. Since such receivers operate strictly on I and Q double sideband bandwidth color information only, phase angles as shown on the vector plot will be shifted somewhat to account for the R−Y (red minus luminance) and B−Y (blue minus luminance) phase shift that occurs so that the receiver's demodulators may operate on a generally quadrature (90-degree) axis. This is normally considered 90 degrees for red and 180 degrees for blue, if the vector reference begins on the horizontal plane at the left, or 90 and 0 degrees, respectively, if the 0 reference is along the horizontal plane to the right. Color sidelock (gaited rainbow) generators normally use the former reference, while older NTSC vector *drawings* use the latter. Tektronix, however, shows zero reference along the yellow side of the diagram in Fig. 5-1 instead of near blue at the right. Therefore, we'll use a single "standard" reference for *all* vector diagrams, be they either NTSC or gaited rainbow, for applied uniformity.

So, with our deliberate move into vectors, let's find out what an NTSC vectorscope is and what it can measure. If you think of this instrument in terms of polar coordinates—that is, some angle and a given magnitude, say 150 ∠28°—you are beginning to understand its function. The mag-

Fig. 5-1. Tektronix's standard vector display of NTSC color bars at 75% amplitude and 100% white reference.

nitude (or radius) of the circular waveform in this instance is simply chroma amplitude, and its angle represents chroma phase. Tektronix suggests that you "think of a vectorscope as an oscilloscope with a circular time base" of from 0 to 360 degrees, with each quadrant representing 90 degrees. Therefore, radii terminus of each swept transition represents a particular color that's present at some specific number of degrees and at a specified distance from other colors which are generated as yellow, cyan, green, magenta, red, and blue—inevitably developed by *any* NTSC generator.

In standard, precision measurements for broadcast of closed circuit TV systems, you may expect the various vector representations to fall precisely into place. With consumer equipment, however, the very nature of its economic design prevents total precision, so we must live with some abnormalities. Nonetheless, a typical TV vector display, when understood, can be just as useful as any phase and amplitude representation, since it does exactly the same thing, at the same time, and in the same way. The very relevant factor that this particular vector pattern is not disturbed by luminance in the picture is an enormous asset.

X-Y oscilloscope instruments (Fig. 5-2) that are used to make such vector measurements also have varying bandpasses, and there will be both phase and amplitude differences from the more precision displays. Therefore, in recording the various waveforms that are subsequently presented, different oscilloscopes such as Tektronix' dual time base,

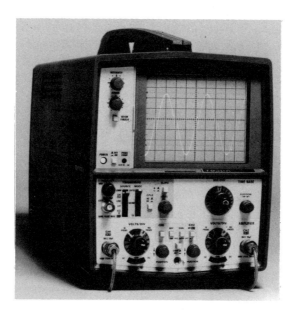

Fig. 5-2. Dual trace T932A Tektronix makes satisfactory lower frequency X-Y measurements. Scope bandpass is greater than 40MHz at 3dB down.

dual-trace T935, Telequipment's similarly equipped D67A, and B & K-Precision's model 1477 dual trace unit all display general waveforms, while the T935/T932 and 1477 record specific vectors (Fig. 5-3). Finally, there is one other difference of new measurement points in modern receivers: Since all signals now go to picture tube cathodes instead of grids, vector waveforms will appear upside down; red and magenta, for instance, emerge at the bottom of the vector rather than the top. The same is true, of course, for gaited rainbow patterns which will follow after the NTSC discussion. If, however, such NTSC vectors were inverted and shown rightside up, then the various voltage levels that produced them would have to be inverted also.

Of course, *voltage levels* are what these NTSC patterns are all about (Fig.5-4). Something like ascending or descending staircases of digital logic, they arrange themselves in step functions so that the three guns of the cathode ray tube are excited at discrete levels, producing pure saturated or combinational colors appearing on the monitor or receiver's picture tube. Two photographs in composite, one above the other but in the same time base setting, prove the point. In trace A, you see the six color bars beginning with yellow and ending with blue. On the cathode ray tube, however, this series actually begins with white—a blending of all three colors in equal intensities—and ends in black, where all three cathode ray tube guns are actually cut off, producing no CRT conduction and, therefore, another non-color, which is black. Below these *eight* analog oscillations, you will see one above the other, green (B), red (C),

Fig. 5-3. B&K-Precisions's 1477 15 MHz 3dB down scope with better than average X-Y performance and lighted graticule.

Chroma and the Vectorscope / 93

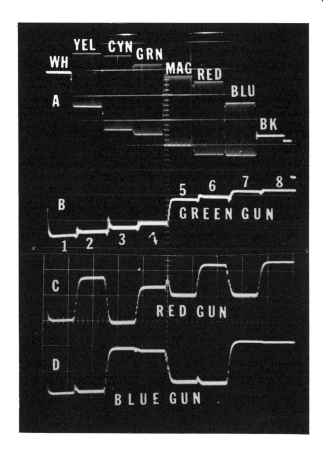

Fig. 5-4. Two polaroid waveform pictures in composite reveal baseband color excitation and resulting green, red, and blue waveforms at the cathodes of their picture tube. The eight steps outlined by the green gun each represent the time interval of either black, white, or an NTSC color.

and blue (D) voltage levels. The most active of these levels is red, and the least active is blue with four sharply stepped functions, followed by green in a series of small steps for yellow, cyan, and green, and close to cutoff for magenta, red, and blue. The three guns of the cathode ray tube are in full conduction when each trace step is down (going toward zero level) and are off at the highest positive level. With CRT grids at AC ground combined with fixed, positive DC levels, higher cathode voltages result in grid cutoff, while lower cathode potentials permit additional beam current, since the grids become more positive when cathodes go further negative. As you can see, cathode instead of grid conduction now controls cathode ray tube beam current—a condition exactly the reverse of outdated CRT biasing, where cathodes were fixed and grids permitted to swing freely with incoming potentials. The reason for cathode biasing is the simultaneous application of DC and AC potentials to three identical RGB electrodes, permitting more uniform wideband video with DC coupling to appear on the CRT phosphor screen. Where there is only grid drive, the color-luminance matrix mix must take place in the cathode ray tube.

THE VECTOR PATTERN

Since the green gun is virtually either "high" or "low" in the NTSC color spectrum, the red and blue guns supply the greatest variations in voltage at CRT cathodes and therefore are used exclusively in generating an NTSC vector pattern. For oscilloscopes with X-Y capabilities, channel 1 is connected via a low capacitance 10X probe to the Red cathode of the picture tube and channel 2 through another 10X LC probe to the Blue cathode. Then, with each vertical channel set to 20V/div (our 10X factor included) the pattern appears as shown if the receiver or monitor brightness, contrast, color, and tint controls are adjusted for a full color, medium brightness picture. Less than full color will reduce pattern size, while misadjusted tint can radically change positions (phases) of the six NTSC colors. Good receivers and monitors will most assuredly produce white for the first bar and cyan for the third bar, with most rendering good magenta and a somewhat brick-shade reproduction for red. Greens and blues on *any* monitor-receiver should be perfect.

Now you may look directly at the NTSC vector pattern (Fig. 5-5)—even though we show no special vector graticule—and see exactly

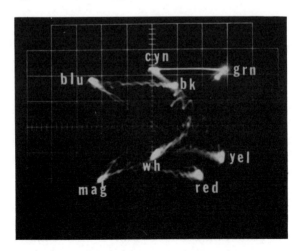

Fig. 5-5. The actual NTSC vector that results from the six color bars plus black and white. Note that all adjacent light points are connected.

where each of these color bars are placed. Remember, we said these would be upside down compared to the initial "baseband" color pattern, and, indeed they are; for red is on the right bottom with magenta, while blue, green, and cyan are at the top. White and black non-colors are almost perpendicular, but the X-Y display won't quite take the rise times necessary to produce straight connections for white-black, or any of the other combinations such as blue-black, blue-yellow, cyan-yellow, or cyan-green, which you will note are all joined by wispy, somewhat indistinct lines in between. Actually, this is how all colors were identified in the first place, since they had to be adjacent to one another to exhibit recognizable connections. The greatest puzzlers were white and black, but since they are placed at 75% and 7.5% on the IRE (Institute of Radio Engineers) scale, respectively, a mental pattern inversion identifies both levels without too much difficulty. All eight voltages now can be positively placed. If your receiver-monitor displays a relatively equivalent 8-point display on a conventional oscilloscope, then you know that it is operating as designed. If not, find the problem by tracing backward through the pre-CRT amplifiers using waveforms shown in the initial composite photo.

If you would like to see the I and Q voltages in the vector pattern also, see Fig. 5-6. They are simply "diagonals" between green and magenta and will probably not be of much use except in very special cases where design or troubleshooting procedures are particularly precise. Note that the remainder of the pattern is not disturbed in either phase or amplitude, so you're safe in using either standard color bars or those with I and Q, if desired.

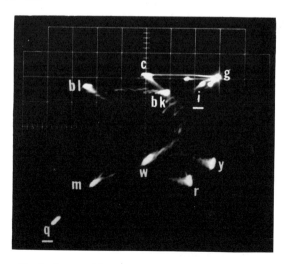

Fig. 5-6. Same illustration as Fig. 5-5, except that *I* and *Q* have been added to show their relative positions.

In summary, NTSC color bar patterns are really the only safe and satisfactory means of testing color circuits in both television monitors and receivers, since pattern generating frequencies are above the customary 30 Hz to 3 MHz luminance ranges and therefore introduce no non-color excitation at all. And for those sets with the new 4 MHz comb filters, color at 3.08 MHz and above is automatically separated from luminance and processed independently, once again excluding luminance from the RGB outputs. When color patterns are transmitted, you will find that an oscilloscope probe connected to the output of the final luminance amplifier produces only a sync pulse with some linear level stepping but no composite video, thereby separating all color functions from the rest of the processing circuits of the receiver.

GAITED RAINBOW COLOR BAR GENERATORS

Although "sidelock" color bar generators are still useful to look at a rainbow pattern of colors from yellow-orange through pink, magenta, blue, cyan, and green on any color cathode ray tube, blanking between bars and seeping luminance in the pattern somewhat destroys their usefulness in RGB (instead of R-Y, B-Y, G-Y) receivers. This is because both luminance and chroma go directly to the cathodes of the picture tube, and superimposed luminance on colors makes the resulting oscilloscope pattern often unusable (Fig. 5-7). However, since DC coupling, DC restoration and chroma-luminance separation are much better engineered in new top-of-the-line color sets, a clean gaited rainbow generator is, once again, becoming quite useful in color analysis and troubleshooting (Sencore CG169, for instance).

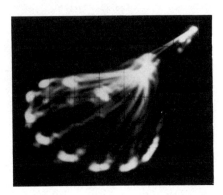

Fig. 5-7. "Parachute" pattern of distorted vector produced in modern red, blue, and green (RGB) color receivers.

Older, grid-driven sets processed chroma and luminance separately until they reached the receiver picture tube, where they were matrixed within the CRT itself. But since direct red, blue, and green drives have become universal, such generators do reasonably well for CRT displays, but not so well oscillographically for earlier 1970 color sets because of added luminance in the pattern. There is no substitute, however, for a *good* 10-bar display on the receiver's picture tube to assess quality rendition of colors.

The "sidelock" generator received its name after someone found that crystal-controlling its frequency exactly at 15,734 Hz below that of the television receiver's 3.579,545 MHz subcarrier oscillator will generate a red, blue, and green raster of color which, when gaited cleanly every 30 degrees, produces 12 color bars (Fig. 5-8). One of these, however, becomes the sync pulse, and a second is blanked by the receiver itself and so you actually see only 10 bars at the CRT terminals. If you can find a way to remove the interfering luminance, then the pattern may well become useful, despite the problem that it is both upside down and horizontally backwards as well.

Numbers denoting 10 petals of the waveform signify this inversion (Fig. 5-9) by being shown in a counterclockwise direction. Were this waveform rightside up and the tint control turned somewhat, you could see that R-Y bar No. 3 should be vertical and that B-Y bar No. 6 would

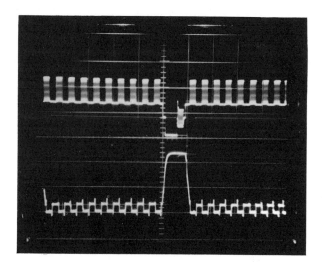

Fig. 5-8. The 12 color bars (3.57 MHz burst) produced by gaited rainbow generator (top) and the resulting 10 color bars that pass throught the receiver. The 10th is blanked by a very wide receiver horizontal blanking pulse, in this instance.

98 / Investigating Video Terminals and Cassettes

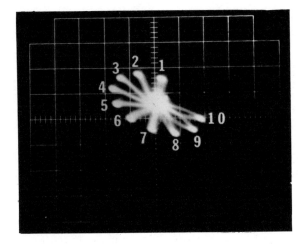

Fig. 5-9. Sidelock vector pattern is both upside down and horizontally reversed. It is still useful, however, in determining chroma processing and demodulation functions. An 80 μF electrolytic capacitor can remove all luminance.

become perpendicular to it. This, of course, signifies full 90-degree quadrature demodulation, which is normally desirable for color processing without too much phase shift combined with picture tubes whose phosphors (rare earth or not) are fairly equal in intensities. X and Z demodulation schemes, which are *less* than quadrature, tend to compress some colors and are often gimmicks to atone for other receiver and picture tube shortcomings. To remove luminance in this instance, we simply connect the positive terminal of an 80 μF electrolytic capacitor to the driver electrode of the luminance amplifier and short any AC to ground, leaving only pure or mixed red, blue, and green excitations. Such unsophisticated luminance short-outs are not always possible in all color receivers and depend very much on circuit configuration. The resulting pattern, then, is the product of a *very clean* sidelock generator and an 80 μF capacitor which wipes out luminance almost entirely. Were these 10 petals just a little more distinct, you would find no crossovers in any petal, a rather uniformly shaped display, and a guarantee that color bandpass and color sync transformers are in excellent alignment. In this instance using the No. 1 petal, the tint control must move said petal at least 30 degrees on either side of the center vertical graticule marker so its range will vary from magenta to green (or cyan). Any color dropout when rotating the tint control shows that the receiver's color reference circuits are improperly adjusted, while petal crossovers positively indicate poor bandpass transformer alignment.

In standard oscilloscope timebase setting at 10 usec/div. and vertical deflection factor attenuation with LC probes at 20 V/div., the R-Y and

B-Y patterns look like those in Fig. 5-10. Observe that the R-Y display (top trace) nulls at the sixth bar (where it should), and the bottom B-Y waveshape nulls at the third and ninth bars, although No. 9 is a little too close to No. 10. This is because the horizontal blanking extends beyond 11 microseconds and unfortunately severs bar No. 10—a design fault indicating sloppy engineering, and indicates poor high voltage and low voltage regulation.

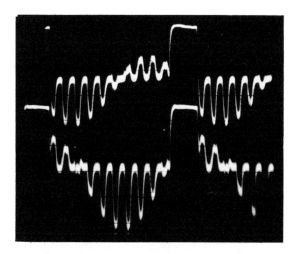

Fig. 5-10. Almost two lines of R-Y (top) and B-Y (bottom) excitation at CRT cathodes from a sidelock color bar generator. Time base setting is 10 usec/div. and vertical attenuator set at 20V/div.

Once again, green does not effectively enter the formation of vector patterns even in the sidelock generator and is not considered. Amplitudes of R-Y and B-Y, therefore, produce the various gaited rainbow color bars seen on the cathode ray tube. In most instances, fortunately, sidelock waveforms at the CRT cathodes are fairly usable even though most rainbow patterns are not. Unfortunately, only the demodulated pattern magnitude is available here—not a vector. Therefore, chroma alignment and proper phase recognition and placement is not available. This end product becomes a *scalar*, a one-dimensional quantity rather than a wholly useful phase and magnitude vector.

CONCLUSIONS

In both color monitor and TV receiver design and servicing, additional signal separations beginning with tuners and extending through horizontal and vertical sync, luminance, and chroma will be required for effective system analysis as the many video terminals become both more compact and more complex. Analog signal levels and digital logic play a much greater part in all portions of any receiver-monitor video display than previously, and will certainly require advanced investigative techniques. Not only is separate tuner, sync, and chroma required, but extra luminance signals to determine overall video response are needed also, (Fig. 5-11.) Fortunately, virtually all such signal sources are reaching the design and service markets now and are just in time as video shifts into high gear for the 1980s. Oscilloscopes, computers, and video terminals of all descriptions are highly intertwined, certainly for the foreseeable future. They're here today and certainly far from "gone" tomorrow.

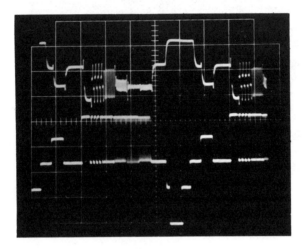

Fig. 5-11. Another 1+ lines of video showing multiburst frequencies between 0.75 MHz and 4.08 MHz (bottom trace) and a rather poor receiver response of hardly 3 MHz at the top. Newer sets with comb filters should show 4 MHz at CRT cathodes.

VIDEO CASSETTES

The video cassette recorder/player is firmly established in the U.S. market as a prime item of consumer electronics, and should become a real bonanza for the service trade as recording heads fail, tapes stretch, sync pulses drop in amplitude, critical circuits develop leakages, and amateur repair attempts foul VCR equipment substantially. In electromechanical machines of even modest complexity, many adjustments and signal-seeking subsystems require more than casual attention, especially in areas of high humidity; and there are considerably more than routine repair fees to be collected in attending such problems.

Video Home System (VHS) units, manufactured by Matsushita, apparently will become the more plentiful since they seem to outsell their rival BETA systems (Sony, Sanyo, Zenith) by as much as 2:1. However, except for recording head configurations, tape wrap, length of play, certain bandpass considerations, and some rather minor modifications in electronics, the two systems are mercifully similar. But while their electronics may be relatively equal, mechanical adjustments are dissimilar enough so that various jigs, plates, tension meters, other tools, and sequential procedures are different, requiring unique approaches to each of the major systems. Allegedly, the BETA types are somewhat more difficult to repair, but technicians adequately trained on both machines seem to get the job done in relatively equivalent times.

From a money-making standpoint these cassette recorder-players and their video disc competition are most assuredly prime consumer servicing objects for the 1980s, and good specialists should profit handsomely. You will find, however, that those technicians and other electronics servicers without adequate equipment as well as sound technical and theoretical instruction, cannot satisfy requirements of critical consumers who have paid upwards of $1000 for these video reproducers in the first place. For instance, is it worth installing a pair of new video heads without certain other waveform checks and adjustments? Of course not! Cassette owners will expect good performance following inevitably substantial bills which complex equipments seem destined to generate.

THEORY OF OPERATION

The BETA system uses an upright head (with adjustments) and 360-degree tape wraparound, while VHS player-recorders feature only a 180-degree tape-head contact with no special head settings but tape-guide adjustments. Since we will be using VHS equipment, predominantly, and to avoid confusion, we'll describe this system's operation primarily, borrowing a few drawings conveniently available from both Quasar and Magnavox (VHS source Matsushita) to illustrate. There was also a 4-head system introduced in 1979 by JVC (Victor of Japan) which is worthy of mention since many later cassette recorders marketed in 1980 will be similar in operation. The four heads do facilitate a useful slow scan and stop mode action with little or no noise in the picture, permitting both 2-and 6-hour recordings on regular tape. JVC stop action takes place primarily during the vertical blanking interval.

Unlike BETA, both VHS tape and tape heads (Fig. 5-12) move about on a tilted cylinder, with tape passing from left to right past a 67 kHz erase head, loading pins, video heads, and three fixed audio heads that are used for erase during secondary or audio dubbing, recording or playback and to record or playback 30 Hz control track pulses. One video

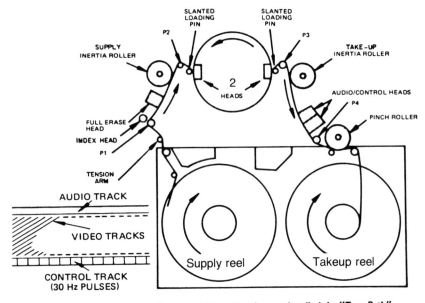

Fig. 5-12. Mechanical portion of VHS system showing tape mechanism and two video heads. (Courtesy of Magnavox)

head traverses a single diagonal field of 262.5 horizontal lines in a single scan and is followed by the other—the total amounting to a standard 525-line frame of which there are 30 per second in each NTSC television picture. At the same time, 30 Hz pulses are either derived from 60 Hz TV sync signals or internally generated to regulate both the 1800 rpm constant speed video head motor and the variable capstan motor.

This motor, designed to drive 1/2-inch magnetic tape, also has a speed sensor frequency generator. The generator delivers 960 Hz in standard play and 480 Hz in long play. It divides down to 240Hz and 30 Hz, respectively for servo and sync circuit control, depending on SP, LP, or SLP 2-, 4-, and 6-hour tape speed selections from the front panel.

Camera or audio input, of course, is *baseband* (no modulation) and bypasses UHF and VHF RF tuners. Tuners, in the better equipments, are all voltage-controlled varactor types with inductor bandswitching shunt diodes and varactor semiconductor backbiased voltage tuning. UHF tuners have, universally, 300-ohm inputs, while most VCRs offer 75-ohm coaxial cable inputs for upper and lower band VHF channels 2 through 13. Many are remote-controlled for channel change, pause, etc., while some of the more advanced units also have slow- and fast-forward motions as well as a stop mode with pictures. All better tuner-timers are microprocessor controlled, select up to four programs each day for some

7-14 days, or repeat a minute or hourly program indefinitely at the same time each day or until the tape runs out. Zenith-Sony also has a new BETA 14-day programmable machine that's just reaching the market in trickle quantities (at this writing) but should have further production in volume before the end of 1980. Ordinarily, a precise time-set clock for day, hours, and minutes tells the microprocessor when to begin programming, (including record duration), and digital logic does the rest.

A metal-covered mylar tape (Fig. 5-13) is magnetized by a video head converting electrical signals into magnetic fields that hold their positions on the tape until obliterated by a full erase head preceding each new recording. With no signal into the recorder, such erasure will assure the simple production of a blank tape, if that's desired.

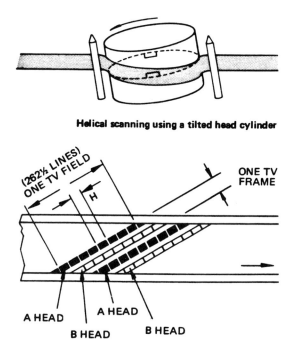

Fig. 5-13. Head slanted, tape moves horizontally as one magnetizes or reads tape during each field.

Tilting each head 6 degrees (Fig. 5-14) in opposite directions minimizes adjacent track pickup during playback, cancelling high frequency interference of one head versus the other in an action called Azimuth Recording. This, of course, is luminance crosstalk cancellation as opposed to chroma spurs cancellation, which when heterodyned between 4.2 MHz and 3.58 MHz to 629 kHz, is now comparatively low frequency; but re-

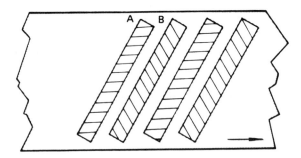

Fig. 5-14. Opposite head tilting—a total of 12°—helps cancel HF luminance interference and is called Azimuth recording. (Courtesy of Magnovax)

mains FM (frequency modulated) just the same as luminance, and traditional TV sound.

To process the chroma signal (Fig. 5-15), a 2.5 MHz voltage-controlled oscillator (VCO) is divided by four and applied to a sequential switch, where phase is rotated at 90-degree intervals between 0 and 270 degrees and is then applied to a balanced modulator supplied by a crystal-controlled 3.58 MHz oscillator. The sum of 3.58 MHz and 629 kHz, of course is the 4.2 MHz. The difference is a 629 kHz chroma downconverted frequency that is processed in the record and play modes for either tape storage or a picture on the TV screen after, of course, recombination with both luminance and sound information. Sound and 30 Hz sync are simultaneously recorded and played from top and bottom *horizontal* tracks, respectively, on the same recording tape. There is also comb filtering of the chroma following the 3.58 MHz bandpass amplifier, using the usual 1h (h stands for horizontal) line delay method for dealing with in-phase and out-of-phase chroma interference elements and ultimately resulting in their cancellation. In the receive mode, each cassette recorder has the usual IF (interfrequency amplifiers) following the tuners and AGC generation, as well as both sound and video detection. All are processed as frequency modulation, with video reconverted to AM for composite signal production in playback to either slave RF or baseband video monitor.

Drawbacks

Probably the biggest drawback to the BETA and VHS systems is their inability to record and playback each other's tapes. Furthermore, some misadjusted recorder-players in either of the two systems cannot interchange because of marginal or incorrect adjustments. It would, therefore, seem reasonable that if black and white levels, oscillators, head equalization, chroma amplitude, and one or two other electronic potentials were reasonably calibrated, at least the signal portion of the BETA

Theory of Operation / 105

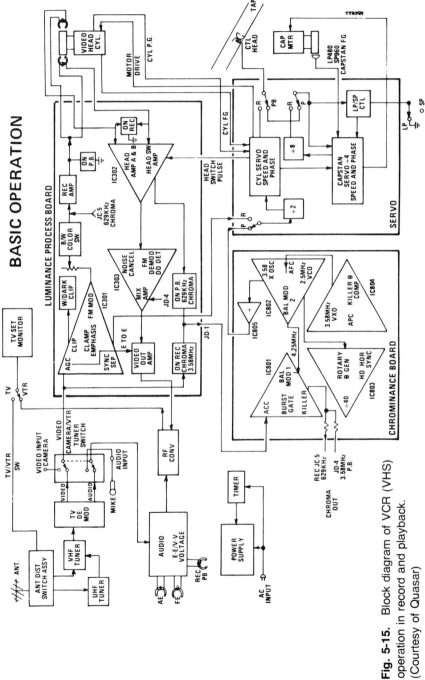

Fig. 5-15. Block diagram of VCR (VHS) operation in record and playback. (Courtesy of Quasar)

or VHS recorders would become equalized and permit favorable interchange within each system, provided tape guides, heads, and tape speeds were properly set up beforehand. After all, most of these tape systems only show about 2-3 MHz luminance resolution (Fig. 5-16), even at 6 dB

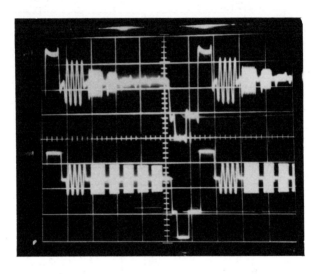

Fig. 5-16. VHS playback 2 MHz display (top) compared with full 4 MHz video multiburst waveform (bottom).

down, in addition to scattered noise and other somewhat undesirable effects—all due, unfortunately, to the interplay of mechanics and electronics in attempting to process some pretty exacting signals ranging from 30 Hz to 4.1 MHz. As you will note in Fig. 5-15, servo (electronic control of mechanical functions) systems are required in both phase and frequency for the head cylinder as well as the capstan, which moves tape in BETA II at 2cm/sec, BETA III at 1.33 cm/sec, and in VHS systems from 3.34 cm/sec in SP, 1.67 cm/sec in LP, and 1/3 of LP for SLP.

Clock Timing

While not important, necessarily, to oscilloscope applications, the digital clock timer (Fig. 5-17) and its associated microprocessor can turn off and on this VCR anywhere up to four times a day for from 7 to 14 days and, therefore, requires some consideration in the overall explanation.

By day, hour, time length, and channel, a random access memory (RAM) is programmed so the recording can begin in some time segment during any day between 0 (today) and 6, which is 7 days hence (14, it is said, in the case of Zenith-Sony BETA III). Four programs may run each of the seven days, or the same program may be taped at the same time

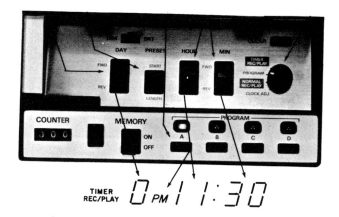

Fig. 5-17. Microprocessor timer/operator controls. (Courtesy of Magnavox)

until the 2-4-6-hr. tape runs out. A central processor (CPU), arithmetic logic unit (ALU), random access memory (RAM), and read only memory (ROM) are all involved in the memory and execution functions, which are manually set by the VCR's operator at times of his own choosing. A/B/C/D program segments per day are then triggered by the clock which starts the microprocessor operation and ends it in time to begin another sequential setting. Whenever the equipment is disconnected or loses AC power, the clock shows an automatic F in place of the day, as well as 12:00 A.M. This means the clock must be reset for additional programming. Channels and their programs may also be blanked by a CLEAR button when changes are desired. Most solid state varactor tuners (voltage controlled) accept any 14 channels—either UHF or VHF—that may be selected. A recent example of a modern player recorder is Zenith's VR9700J BETAMAX (Fig. 5-18), which records up to four programs on four separate channels during four different periods over any 14 consecutive days. The clock set at five minutes after midnight is self-evident, as is the cable-connected remote control, featured by all current deluxe units.

Troubles and Pertinent Waveforms

There are problems with these equipments, especially since they are electromechanical and must have periodic maintenance and adjustment for anything better than marginal or actually inferior operation. You will quickly discover, as shown in Fig. 5-19, that modulated and unmodulated staircases are not often linear (denoting differential phase and gain errors), and overshoots and even preshoots have a habit of cropping up in

108 / Investigating Video Terminals and Cassettes

Fig. 5-18. The Zenith-Sony BETAMAX VR9700J 14-day recorder with remote control.

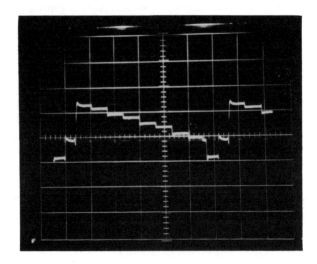

Fig. 5-19. A nonlinear unmodulated staircase checks some very useful luminance discrepancies.

strange places, presenting, occasionally, some undesirable transitional effects. You will also discover that standard colors at baseband out of the NTSC generator will not appear as symmetrical mirror images after passing through the VCR, RF, IF, and video detectors of even wideband, synchronous-detecting deluxe color sets, as the top waveform (VCR) versus the bottom waveform (baseband) is compared in Fig. 5-20. Were you to take VITS and VIRS off the air, encompassing vertical blanking interval lines 17 through 19, you'd find nothing like the voltage representation illustrated in Fig. 5-21. Unfortunately, VIRS wouldn't be very usable, certainly not until these player-recorders expand their luminance bandwidths to at least 3 MHz and have improved considerably the bandpass of 629 kHz chroma. VITS, however, is another matter, and both NTSC color patterns and multiburst will prove to be the most reliable design and troubleshooting aids available. Luminance bandpass, by the way, drops substantially following amplification and clipping in the first luminance IC processor.

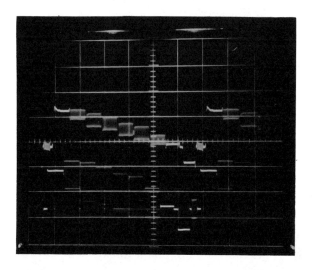

Fig. 5-20. Comparison of color bars passing through VHS recorder with baseband out of the testing instrument (bottom waveform).

Here are a few pointers to remember:

- With head cylinder or capstan cylinder out of sync, you'll immediately hear low-frequency audible noises.
- Hum bars are evident in heavily worn or peeling tapes.
- Head problems present steady or up-moving visual interference.
- Tape problems are either steady state or move down.
- Color breakup is often due to improper head switching.
- Always check LP against SP, head switching, capstan servo sampling, and cylinder servo sampling. Occasionally, the free run oscillator with and without signal needs touchup also.

Here is a point you should *never* forget: Don't ever try and "diddle" these VCR's by screwdriver and eyeball. There is no way under any

110 / Investigating Video Terminals and Cassettes

circumstances that a misadjusted machine will ever play more than the single tape on which you have managed to somehow inscribe a recording. When manufacturers' instructions say 0.5V ± 0.1V, they mean exactly what they say. First, you must have excellent equipment, such as 3 percent oscilloscopes and 0.05 percent (DC) 4 1/2 digital multimeters, and 7-place accurate counters, or look for another business. The cylinder servo sampling gate adjustment is an excellent example (Fig. 5-22): either these two waveforms—one with extremely fast rise and fall times—are synchronized or your equipment is running randomly out of sync.

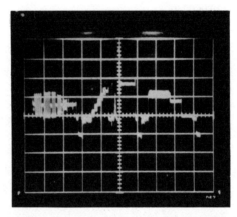

Fig. 5-21. What a good oscilloscope can pick up directly from the air, showing lines 17 through 19 of the vertical blanking interval. Both VITS and VIRS are evident.

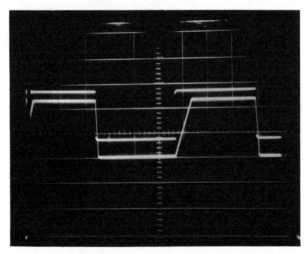

Fig. 5-22. If you think scopes shouldn't be precise, look at the risetimes in this cylinder servo display. For perfect sync, the timing of these two waveforms must coincide exactly.

Theory of Operation / 111

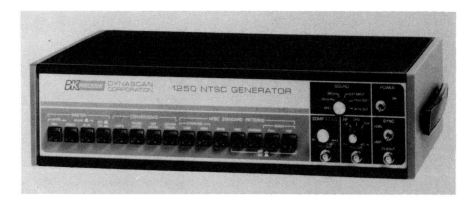

Fig. 5-23. B&K-Precision Mods. 1248 and 1250 gaited rainbow and NTSC generators used in these signal investigations. The 1250 has great flexibility, including external video and chroma inputs that can be RF modulated.

Chapter 6
Waveform Analysis

One of the most difficult problems for either electronics professionals or journeymen is the ability to recognize faulty waveforms. Some engineers and technicians spend their entire careers trying to interpret subtleties among certain voltages and currents (with transducers) and still don't understand what the fundamental procedure is all about. This chapter will investigate thoroughly a commonsense fault analysis and produce working procedures, coupled with adequate background information, that should become a boon to all those who still think waveforms are some sort of purple or pink magic. Especially noteworthy in the "tradesman" industries are the two-way radio and TV personnel who either don't use oscilloscopes at all or who are attempting to use them with woefully inadequate expertise and virtually no primary instruction. Under such circumstances, it's no wonder that service shop failures are multiplying faster today than at almost any time in the past. No doubt about it, modern electronics are tough! And there is no reason to believe they will ever regress and simplify.

BASIC CONFIGURATIONS

Just as Latin students must remember that ancient Gaul was forever divided *in partes tres*, modern electronics scholars should know that basic waveforms tracing the flow of electrons are sine waves and that any periodic nonsinusoidal voltage or current consists of some initial sine wave and additional sine waves (or harmonics) that are whole multiples of the original. These combinations could be either pulses or square (rectangular) waves, with duty cycles less than (<), equal to (≡), or greater than (>) 50% (Fig. 6-1). From such waveshapes one can derive triangles, ramps, integrated, differentiated, clipped, clamped, and many other representations that are common to electronic processing. It's not the simple generation of these voltages and currents that is of primary interest, however, but what they do and what is done to them. The basic journalistic queries of "who, why, what, when, and how" still apply to much more than the print and picture media. The *significance* of any electronic function continues to be of prime importance, along with the type of drive impetus that makes it operate.

Basic Configurations / 113

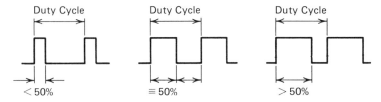

Fig. 6-1. Complex waveforms consisting of a fundamental sine wave and harmonics form displays of varying duty cycles.

Square (or Rectangular) Wave

The square (or rectangular) wave then, becomes the subject of our initial investigation because *both* high and low frequencies are involved and service problems more simply diagnosed (Fig. 6-2). As many already know, any square wave is composed of a fundamental sine wave and its odd harmonics; the more numerous the harmonics, the squarer the waveform. Since harmonics tend to bunch together during rise and fall times, their shorter durations constitute high frequencies, and their longer, bumpier characteristics appear as low frequencies on the tops (and bottoms) of these same displays. Further, it doesn't make any difference whether such rectangular voltages are generated by analog means or digital switching; they all have the same harmonic content. What is good for one applies explicitly to the other, and this is the real reason that the square wave family is so useful in both digital and analog electronics. It checks both high and low frequencies at a glance, telling you what your circuit, discrete transistor, or IC problem might be and readily supplying a solution.

For instance: in (a) and (b) of Fig. 6-3 you see preshoot and then overshoot, usually some small capacitative or RC effect; in (c) there is damped ringing, normally induced by an inductor; in (d) high frequency loss, this time by another small capacitor; (e) denotes phase shift, which is often an RLC (resistor-capacitor-inductor) effect; (f) you may often see in low-frequency rectangular waveforms that are capacitatively coupled; (g)

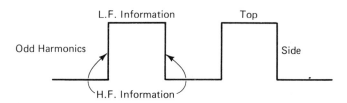

Fig. 6-2. Rectangular waves contain both high and low frequency information. Harmonics are odd (3, 5, 7, etc.)

114 / Waveform Analysis

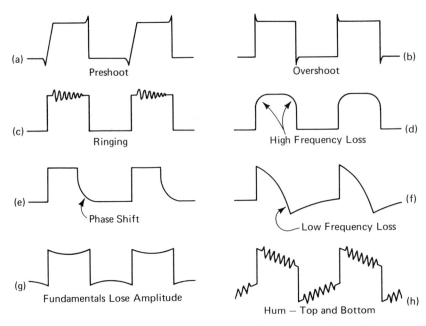

Fig. 6-3. Various "square" waves showing symptomatic problems.

where the fundamental sine wave itself is losing amplitude from any number of causes, including low voltage from the power supply; and (h), where lower frequency information appears as hum on the waveform's top and bottom. Were the frequency faster than the rectangular representation, obvious interference would have been a smear instead of plain hum.

By applying the lessons of these eight waveforms, almost any coupling, bypass, or filter problem can be found in both digital and analog pulse-type circuits. If you drive digital square waves into regular analog circuits, their rapid rise times and harmonically rich flattops will tell you a great deal about basic analog design and subsequent operation. On the other hand, lower frequency sinusoidal voltages impressed on digital rectangular waves will look like some sort of strange and undesirable modulation.

Triangular Waves

Triangular waves often combine desirable testing qualities contained in both sine and square waves and will reveal clipping, crossover distortion, gain changes, and high-frequency rolloff as well as low-frequency distortion—all of which you see contained in Fig. 6-4. The clipping (top and bottom) in (a) is obvious and occurs when some analog circuit is not permitted its full voltage swing. Crossover distortion (b) appears most often in push-pull output amplifiers and should be smooth both for linear-

Basic configurations / 115

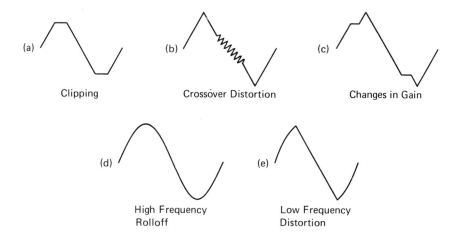

Fig. 6-4. Triangular waves readily find specific types of distortion.

ity and maximum transfer of power. Gain changes, as illustrated in (c), often materialize in amplifiers because of improper biasing. Both low-frequency and high-frequency rolloff and distortion (d,e) are normally LC induced, and their causes should become quickly apparent when any circuit design is evaluated.

Sawtooth Waveshapes

Sawtooth waveshapes (Fig. 6-5) consist of a fundamental and all odd and even harmonics present. Not normally used for testing, such voltages and currents are circuit and stage drivers and are often found in oscilloscope time bases, deflection amplifiers, picture monitors, and almost anywhere ramps are needed. Usually the up-ramp is relatively slow compared with its more rapid off-time (which usually denotes a blanking period), permitting the forward trace to "fly back" to its original starting position and begin its ramp ascent once more. Naturally, the more harmonics the steeper and more linear the trace. Sawtooths of voltage, however, don't produce sawtooths of current in inductors, and can be substantially misshapen by wrong-value capacitors. It takes a rectangular wave of voltage to produce a sawtooth of current in inductors driven by solid state devices. Tubes driving inductors require trapezoidal (rectangular and sawtooth) voltages because of associated higher impedances. Therefore, even though sawtooth displays are not necessarily perfect

Fig. 6-5. Sawtooths are normally *not* used for testing.

testing solutions, they do have important functions as driving and timing devices and require both linearity and amplitude attention if they are to fulfill any and all compatible missions. It's well to remember that many, if not most, sawtooths are formed by charging a capacitor and using the linear portion of its charge (or discharge) for a positive or negative ramp. Naturally, changes in value or capacitor leakage can destroy the linearity of any sawtooth as well as its on and off times.

Staircases

Staircases (Fig. 6-6) are also useful sawtooth-type evaluators when each step has the same dimensions as the others and is (usually) monotonically ascending. Good for video gray scale evaluation, luminance distortion, differential gain and phase, dynamic gain, etc., such staircases may be either modulated or unmodulated, depending on specific measurements required.

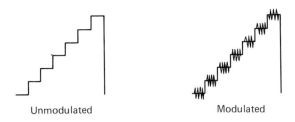

Fig. 6-6. Staircases, both modulated and unmodulated, check linearity and gray scale.

When modulated, stairsteps are handy in analyzing luminance nonlinear distortion and will also indicate color subcarrier changes versus luminance changes—a distortion measurement known as differential phase. To test linearity, a ramp of 5-step, or 10-step staircases are ordinarily provided in the better equipments. For video examinations, 10 levels (steps) of grayscale is considered standard. Naturally, you'll have to be in video work of some description to use such waveforms, but they're enormously valuable for specific evaluations.

In addition to staircases, there are two additional video signals that are of enormous use to broadcasters now and will become equally important to video cassette recorder-player and general home television equipment later. They are called VITS and VIRS. With these two displays, just about any luminance or chroma (color) condition can be verified or diagnosed; consequently, they are highly important in any monitor receiver analysis.

VITS

VITS, the more important of the two color reference signals, stands for *vertical interval test signal* and is required daily-nightly transmission for any TV transmitters remote from their studios. VITS consists of fields 1 and 2, line 17 of the vertical blanking interval (21 lines in all) and field 1 of line 18 (Fig. 6-7). The signal contains a horizontal sync pulse and six

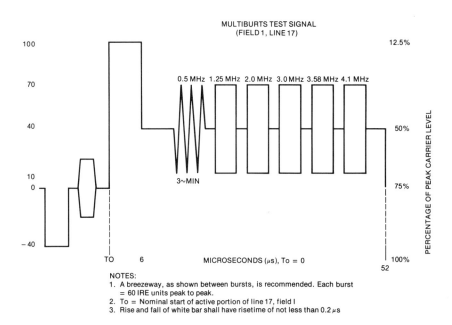

Fig. 6-7. Multiburst portion of VITS seen on field 1 of line 17 during the vertical blanking interval.

distinct multiburst sections of 3 cycles minimum, ranging from 0.5 MHz to 4.1 MHz on line 17, field 1; a horizontal sync pulse, color burst, and a full 8-bar color test signal, including white and black as well as yellow, cyan, green, magenta, red, and blue on line 17, field 2 (Fig. 6-8), and a composite display (Fig. 6-9) showing the horizontal sync pulse and burst (once more, since this is another whole horizontal line), a 6-step modulated staircase, 2T and 12.5T $\sin^2$ pulses, and an 18-microsecond window—all referenced in terms of a -40 to $+100$ IRE (Institute of Radio Engineers) scale that is normally expected to measure 1 volt per 140 IRE units. Since only special monitor oscilloscopes will reliably sync each line of the vertical blanking interval, ordinary engineering and service-type oscilloscopes normally show the entire VITS broadcast dis-

118 / Waveform Analysis

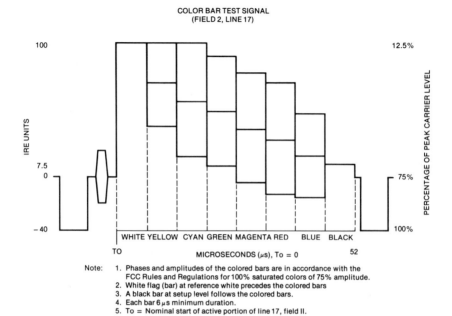

Fig. 6-8. The six color bars plus black and white on the chroma test signal, which appears on field 2, line 17.

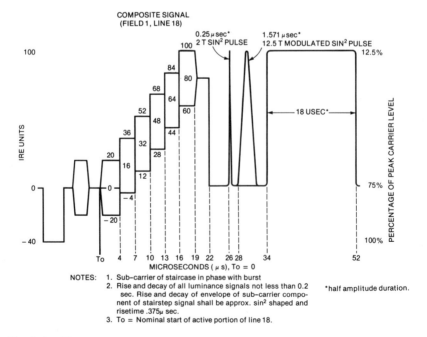

Fig. 6-9. The composite signal consisting of staircase, 2T and 12.5T sin² pulses, and the 18 usec window on field 1 of line 18.

play of three fields on two lines, if you're lucky. Otherwise, you'll have to purchase a B & K-Precision 1250 for color bars and staircase, or a considerably more expensive Tektronix composite sync and NTSC generator to supply the works as a synthesized signal, plus just a little bit more. A single, continuous intelligence generation into your monitor or receiver is far more preferable in terms of accuracy and reliability, but such instruments producing said information are still quite expensive. However, if you're in the video business—whether broadcasting or home terminals—it's very likely that sooner or later you will become a proud owner. Now, let's see what VITS actually does.

If put through a broadcast system, video monitor, or home receiver, VITS tells you immediately if your system has a bandpass that extends from 0.5 MHz to 4.1 MHz and how linear the system is. For instance, if 3.58 MHz and 4.1 MHz multibursts were little more than a narrow, straight line (which they are not) and the 3 MHz oscillation was within 6 dB of being normal, you could consider that your system had a fair 3 MHz bandwidth (Fig. 6-10). Such a passband would be all right in 1980 for a monochrome monitor, but it should extend to at least 4 MHz for either a good color monitor or home receiver. With the arrival of precise chroma-luminance separation by comb filters, there's no reason that 4 MHz of horizontal scanning resolutions (about 330 lines) shouldn't be plainly seen. If these multiburst oscillations don't follow one another laterally (but not necessarily at the same amplitude, however desirable this may be), you will find that the tuned circuits through which they have passed are out of alignment. However, aligning tuned circuits such as video IF

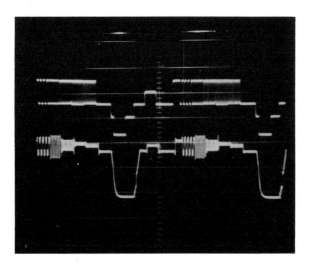

Fig. 6-10. With multiburst from 0.75 MHz to 4.08 MHz showing in top waveform, a *standard* color receiver will display usually no more than 3 MHz at its cathode ray tube.

(intermediate-frequency) amplifiers should be done with a signal source, such as offered in Tektronix or Sencore's VA48 MATV-VIDEO analyzer, rather than trying to work with station transmissions off the air. The broadcast signal, unfortunately, may or may not offer total reliability for a number of reasons, including the state of your own antenna and lead-in transmission line.

The color bar test signal in field 2 of line 17 displays the six colors as well as non-colors, black and white, along with the usual horizontal sync pulse and color burst. It is mainly useful for relative amplitude checks, color placement, and any top and bottom transients or ringing that would indicate nonlinearity. It isn't always seen on broadcasts, as we will demonstrate shortly.

The *composite signal* appearing in field 1, line 18, however, is highly useful both in the studio, for monitors, and in home video terminals since its functions are legion. As already suggested, modulated and unmodulated staircases are very helpful in checking both luminance and chroma relative distortion as well as ramp and phase linearity. The 2T and 12.5T pulses that follow, however, have special purposes too. The 2T sine^2 pulse is a prime operator between 1 and 3 MHz where it is especially useful for frequency response observations as well as group envelope delay. When it has the same amplitude as the window following, the frequency response passing through tuned amplifier circuits is peaked to at least 3 MHz. This 2T test pulse is unmodulated.

Its 12.5T companion pulse, however, *is* modulated, and covers frequencies primarily between 3 and 4 MHz, which involve much more chroma. Once more, amplitude relative to the window height is important, just like its baseline. An uneven baseline denotes chroma delay as well as chroma loss through tuned circuits. If red lettering on white backgrounds, for instance, overlaps toward the right, this "bleeding" represents chroma delay and is often obvious in many older video equipments.

The 18 usec bar pattern is used to measure line and field time distortion, perhaps high frequency rolloff, and most assuredly any type of ringing that can appear in either monitors or broadcast equipment. Top or bottom tilts of this window denote field time distortion. Tilts of the horizontal sync pulse show low-frequency rolloff. Compression of burst, following this sync pulse, can affect color amplitude in receivers, while a change in its phase means a shift in hue or tint.

VIRS

The vertical interval (color) reference signal (VIRS) is, once more, equally applicable to broadcasters as well as to monitors and receivers (Fig. 6-11). Unfortunately, this signal as yet has no Federal requirement

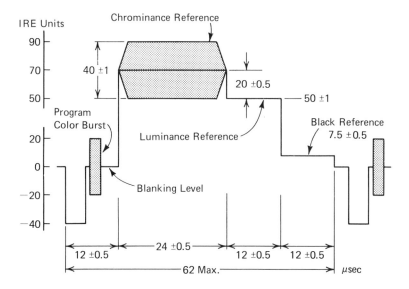

Fig. 6-11. The highly useful VIRS display—good for both studio and some video cassette chroma phase and amplitude adjustments.

and is placed on round-robin broadcast networks and rebroadcast by local TV stations as an *aid* to chroma phase and amplitude control but not necessarily as a universal standard. Television receivers—both direct view and projection sets—are now making considerable use of the VIRS properties for chroma phase and amplitude regulation, and the receiver industry is expected shortly to adopt VIRS as a universal means of color control, which should require its mandatory broadcasting.

VIRS is not an especially complex signal, but the results of its effects can actually control all of the color broadcast. Actually it is a "diagnostic test signal," according to the Electronic Industries Association (EIA), "intended primarily to monitor and measure the characteristics of a transmission device or facility and is not associated with any individual program." Inserted on both fields of line 19 in the vertical blanking interval, VIRS has been formulated to represent "typical program levels rather than extreme or peak values" and it consists of a 24-microsecond chroma reference of 40 IRE units superimposed on a luminance pedestal of 70 IRE units. The phase of this reference must be the same as burst, found on the back porch of the vertical sync pulse.

After the chroma reference is a microsecond luminance reference horizontal bar lasting for 12 microseconds at 50 IRE units, followed by a 12 microsecond black reference at 7.5 IRE units. Placing luminance reference in contact with chroma reference allows an instantaneous check of chroma/luminance ratio. During broadcasts, the luminance level is first fixed, followed by black, chroma, and burst levels in reference to it. Thus, the VIR signal should correspond to the program signal into which it has been inserted. In good quality video tape recorder applications, for instance, a VIRS input could allow playbacks to be properly adjusted for adequate and consistent reproductions of tape programs or commercials wherever distributed or broadcast. When burst and chroma reference phase are not exact, a vectorscope dot will show directly above color burst. In other instances, the 9 cycles of burst may be directly compared in alternate sweep instruments with chroma so that they are operating together.

As described, then, industrial or broadcast video recorders should be capable of receiving and processing the VIR signal without undue distortion. If they can't, then adjustments are required within their limitations. In this way, one recorder can be made to accept tapes of another and reproduce an acceptable image from the second recorder-player source, provided that the mechanisms and tape speeds are similar. VIRS-VITS generators as either auxiliary or dedicated instruments are now available from Tektronix and should be on the service market by other test equipment manufacturers by the time this book is in print. When combined with NTSC color bars, staircase, and a good modulator, this VIRS-VITS combination can pretty well analyze whatever you want to know about any signal-processing video terminal. For help in checking Teletext and Vuedata, signals of these types might also be included in the more deluxe equipment (Line 21, of course).

Composite VITS-VIRS signals are normally available off the air for those with oscilloscopes having dual time bases and exceptional triggering. Normally, however, these scopes haven't the ability to look at individual horizontal scanning lines, especially with complex information, so they often "double up" and show several lines at a time. Consequently, in Fig. 6-12, the VITS-VIRS illustrated covers the entire broadcast transmission now available between lines 17 and 19 in the vertical blanking interval. All the functions shown will be remembered from previously drawn diagrams. An exception is the composite video (color bars) portion which apparently was not included in the broadcast intelligence, or the scope, itself, failed to pick up field 2 of line 17. Sometimes, also, color bars and multiburst will alternate (flip-flop) on the scope's graticule as they try to trigger on the analog information.

ACTUAL WAVEFORMS

Let us now turn to real displays where actual circuits are forming waveshapes that our oscilloscope "sees" and presents for your inspection. How well the displays resolve, of course, depends on the ability of the oscilloscope to "handle" the various fast and slow inputs (triggering), trace brightness (accelerating voltage), and amplitudes (bandwidth and

Actual Waveforms / 123

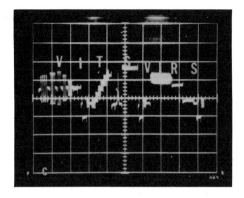

Fig. 6-12. VITS and VIRS appearing in the same waveform from a broadcast station.

volts/division). Here we're using a Telequipment (Tektronix) D67A, that is 3 dB down at 27 MHz, with 10 kV accelerating potential, excellent spot size, dual trace, and dual time base.

To get the feel of one of the "how's" in waveform measurements, consider the X-Y pattern in Fig. 6-13. Although normally representative of *sharp* diode knee characteristics as seen on a curve tracer, this display was actually contrived by putting a function generator into one vertical Y1 amplifier and using simple sweep to deflect the Y2 of the D67A oscilloscope. It neatly demonstrates the operation of the scope's electron beam that is controlled by static deflection plates and their horizontal/

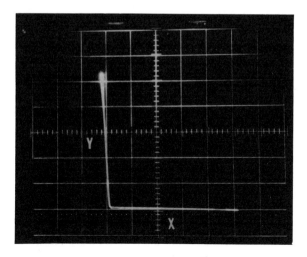

Fig. 6-13. A sharp diode knee represents the Y vertical and X horizontal axes of any ordinary laboratory or service oscilloscope.

vertical amplifiers, which are calibrated in millivolts and volts/div. (Y) and seconds, milliseconds (10^{-3}), or microseconds (10^{-6}) per division (X). The result is a Y = (f)X function, producing a continuous readout in two dimensions.

Since this particular scope has two time bases, a good example of mixed sweep is decidedly in order so that you may use it when examining certain portions of any waveform whose repetition rate is not outrageously excessive. By this, we mean nanoseconds or low microseconds for the medium frequency bandpass oscilloscopes. In similar oscilloscopes, a second time-base generator called B is added so that a portion of the sweep normally carried by A can be delayed. Any portion of the sweep diverted to B may then be intensified and control-selected by the delay so that the precise area to be investigated is plainly seen. This area may be peeled off (mixed) and seen as an expanded portion of the original signal with all initial information remaining, or the intensified portion can be separated completely from the general display and viewed independently as a separate and distinct waveform. For millisecond and microsecond signals, such time base addition offers the very considerable advantage of being able to separate otherwise non-viewable information and accurately "read" both time and amplitude as well as permitting a thorough search for any performance-limiting interference or transients that may or may not appear.

Now that you know what oscilloscopes do and how they do it, it's time to study the various kinds of waveforms found in ordinary measurements, and review their symptomatic faults and peculiarities. Understanding the origin and application of the many "standard" ac signals helps future design and troubleshooting to appear considerably more simple in both systems and individual circuits. After all, current through any impedance either charges that passive element or develops a voltage output across its collective reactance and/or internal resistance. Remember, also, that collector, drain, or anode current in active devices is the output of any semiconductor, which is normally *current*-operated, while the transconductance of any grid-operated tube becomes wholly dependent on grid *voltage* and varying changes in plate current and is known, therefore, as a voltage-operated device. Both tubes and semiconductors, nonetheless, deliver anode current, which inevitably develops into voltage signals that may be "read" by any suitable oscilloscope.

Up to some point, higher reactance/resistance, dc operating voltage and stage gain all contribute to the amplitude of any waveform. Afterwards, it is up to the technician or engineer to determine this waveform's linearity, origin, processing, and effectiveness. Our problem is to discuss analog and digital waveshapes analytically and effectively, and yours is to learn the various signs of good and bad electrical-electronic signals. Once you have a "feel" for system and circuit operations, your electronic abilities will be substantially improved.

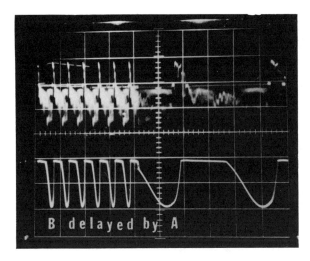

Fig. 6-14. Mixed sweep (B delayed by A) is illustrated for both Y amplifier channels.

Fundamental Waveshapes

Fundamental waveshapes are products of resistors, capacitors, and inductors, either singly or in combination. Overall impedances often include resistors, along with other reactive elements, since they will change both signal magnitudes as well as phase directions. As you should know, however, only a pure resistance passes both current and voltage in identical phase. Now, since there is loss, never amplification, in discrete (individual) passive components, various active devices such as transistors and vacuum tubes must be included in any system for rectification, switching, and amplification. Consequently, both passive and active components have to operate effectively in such subsystems, and linearities or switching speeds depend on careful component placement and use. Taking all these factors into collective account, let's consider fundamental waveforms and then work up.

Rectangular Voltages

Rectangular waves are first (Fig. 6-15). This is because a collection of sine waves and their harmonics can form rectangular waveshapes of varying duty cycles, usually beginning with the well-known square wave that is on for 50% of the time and off for 50%. The swing of such a cycle, of course, approximates that of zero (0) to some specific dc voltage (less energy dissipated in the load), which could well represent the power supply; consequently, it's used as a universal calibration source for many waveshaping and evaluating instruments. Every oscilloscope, for instance, normally contains some fixed 500 mV to 2 V, 60-1,000Hz square wave calibration source.

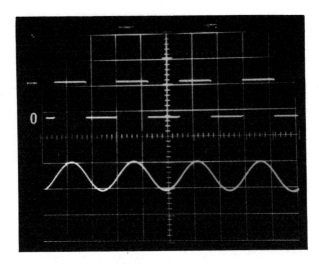

Fig. 6-15. A square wave is formed from many sine waves. The "0" indicates where the square wave begins to swing toward some positive voltage and cuts off at the top of its swing.

In any rectangular wave, time of duration, rise and fall times, and symmetry are the more important considerations. At lower frequencies, or repetition rates, a sharply rising leading edge followed by a long fall time could well indicate capacitance coupling. At very high frequencies, inductive coupling or shunts might well result in ringing (Fig. 6-16), which in any video display can be seen as a series of light or dark stripes, depending on drive and oscillating frequency. When both rise and fall

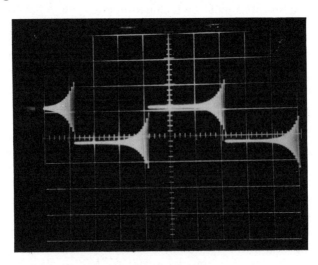

Fig. 6-16. Several cycles of ringing demonstrate quickly damped oscillations.

times are sloppy, linking impendances are often at fault, especially those with larger than normal capacitance.

Other waveform oddities among pulses and and square waves are conditions of preshoot or overshoot, which follow one another when transitions change. In Fig. 6-17, you see a square wave with both overshoot and symmetrical glitches on the upside and downside of the waveform. More often than not such effects come from inductor-capacitor combinations, especially intermediate frequency coils, where current will lag the impressed voltage by 90 degrees. But since current in capacitors leads voltage by 90 degrees, you will have crossover charge-discharge waveform portions resulting in obviously visible transients. In other instances, preshoot can be imposed deliberately or inadvertently on analog or digital voltages for certain effects such as generating a small trigger pulse or for highlighting dark-to-light transitions, making them more apparent. If, however, such transients as those in Fig. 6-17 do materialize in ordinary digital logic, then you may well have false triggering or delays that will undoubtedly upset your pulse train(s). Obviously, your problems are in the higher frequencies here since disturbances are on the fast rise and fall times rather than low frequency response along the top.

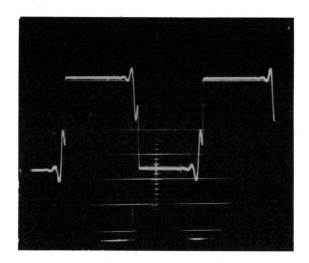

Fig. 6-17. Overshoots and glitches.

When evaluating pulses or rectangular voltages, the arithmetic inverse of one symmetrical cycle amounts to frequency, and a single cycle represents the distance from rise or fall transition to the second respective rise or fall transition. Pulse (or sync) repetition rates may be found by simple equation: Pulse repetition frequency PRF = 1/TP (time between pulses); or simply measure from start-to-start or finish-to-finish on the 10 vertically striped oscilloscope graticule, adjusting the time base so there are convenient spearations between pulses. Otherwise, measure from waveform tip-to-tip and invert the result, letting the reciprocal become your answer. As in Fig. 6-18, if the time base is set at 50 usec/div., calculate the repetition rate. How about 3.45×10^3, or 3.45kHz? ($5.8 \times 50 \times 10^{-6} = 1/(290 \times 10^{-6} = 3.45 \times 10^3)$

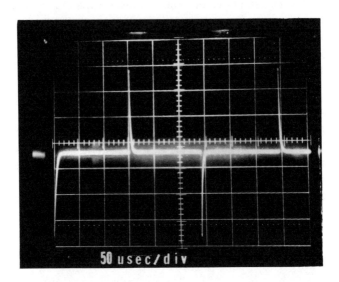

Fig. 6-18. A differentiated display is shown at 50 usec/division.

Sinusoidal Voltages

Sinusoidal voltages are second. Sine waves (Fig. 6-19), as they are now called, answer to the basic equation

$$e = Em \sin \omega t$$

where e is the instantaneous value, Em the maximum voltage, and $\sin \omega t$, the sine of $2\pi f$ (6.28 × frequency), which determines placement of the sine wave rotation times its repetitive frequency.

$$i = Em \sin \omega t$$

is the electron flow voltage equivalent, while power becomes

$$P = i^2 R, \text{ or } RIm^2\sin^2\omega t$$

with capacitive or inductive phase angles amounting to

$$X_L/R = \tan^{-1} \text{ (arc tangent)}.$$

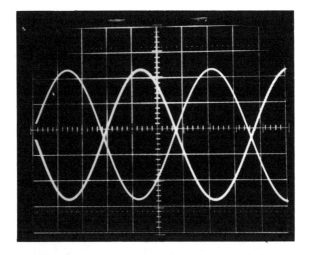

Fig. 6-19. Sine waves 180 out-of-phase joined together.

All the above are equally applicable to both series and parallel circuits once their pure resistances and resultant vector reactances have been established. The real importance of sine waves is that any type of wave form that's nonsinusoidal can be reduced to various voltage or current combinations which are actually sinusoidal except, of course, dc. Therefore, a pure sine wave (really the father and mother of all displays) is irreducible and consists of but a single frequency component, itself!

While a number of sine waves, or sine waves and their harmonics, are very useful because they form additional waveshapes, the fundamental sine wave is as much a child of electricity as electronics. In any application, its configuration is quite important, since this is the only waveform that shapes voltages and currents identically. Consequently, when examining characteristics of any analog network, stage, or circuit, it's common to assume the input is a sine wave so that currents and voltages are treated and calculated accordingly. Sine wave reshaping and modification immediately indicates the addition of more frequency components not present in the original waveform—a valuable factor in analyzing poor analog circuit performance among RF carriers, tone bursts, multiburst, etc., as well as low-frequency audio. Sine waves, however, are *not* as sensitive to minor waveform discrepancies as are rectangular voltages, therefore harmonic-laden test signals are almost more useful than sine waves in most applications.

Phase shifting or distortion, however, is another matter entirely, and here is where sine waves stand tall. But there must be two or more for comparison (Fig. 6-20) since solitary sinusoidals exhibit no visual change with only fundamentals present. Therefore, phase delays, frequency, amplitude, the introduction of harmonics, and other similar irregularities may be studied with advantageous results. This is especially true for frequency differences and sine waves at quadrature (90° out of phase). Any Lissajous figures (X-Y scope inputs) will also determine frequency if one reference is know. A circle, for instance, results, from two voltages 90° out of phase, but both must have the same frequency and amplitude. On the other hand, if you have three complete cycles of some waveform and the addition of another increases these to 5 (Fig. 6-21), then the ratio

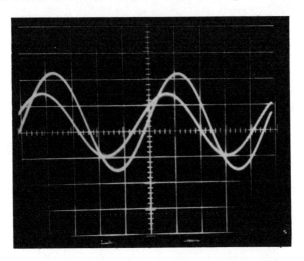

Fig. 6-20. Sine waves of neither identical amplitudes nor phase superimposed.

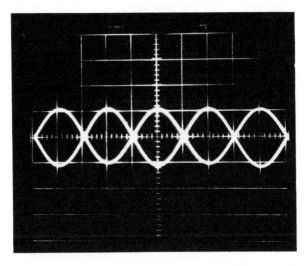

Fig. 6-21. A 5/3 ratio of the initial waveform would be a single cycle.

is 5/3 times the frequency of the initial waveform. Were the controlled frequency 2.5 kHz, for example, then

$$F/2{,}500 = 5/3 \text{ or } 4166.67 \text{ Hz}$$

which is the value of the unknown frequency. In a simpler, related method, the number of loops side-by-side (horizontal) or vertically superimposed denote the frequency ratio: 2:1, 3:1, etc.

Sawtooth or Ramp Waveforms

These usually consist of odd and even sinusoidal harmonics, or just the output from the linear waveform portion of a charging capacitor. They may be either positive or negative going (Fig. 6-22) and in many instances quite linear, with voltage and current changing proportionally. Their retrace time, however, is not linear and is often blanked during stop-start trace intervals. Unlike square waves which are formed by only the odd harmonics and cos θ, sawtooths involve both even and odd harmonics and are predicated on the sine of the angle theta (θ).

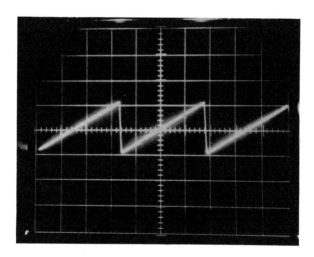

Fig. 6-22. A ramp or sawtooth, whose fall time is much faster than its rise.

Ordinarily, sawtooth energy is used for driving other circuits rather than testing. But when sine and square waves are combined in a triangular wave (Fig. 6-23) that operates on the 9th, 25th, and 49th, etc. harmonics, they will easily show clipping, crossover distortion, high-frequency rolloff, gain changes, and low-frequency distortion. However, the sawtooth is very useful in oscilloscope sweeps, TV output sync driving, AFC phase comparisons, et al., and eminently suitable for any ramp type generation as long as the trace portion is linear.

132 / Waveform Analysis

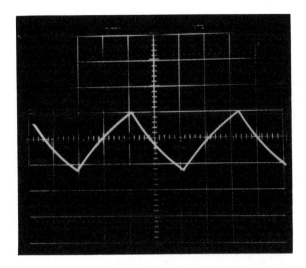

Fig. 6-23. Sine and square waves form triangular waves useful in illustrating clipping, hf rolloff, crossover distortion, etc.

Composite Video Waveforms

Composite video waveforms consist of two varieties, each dependent on an oscillscope's time base setting for adequate viewing. At 59.94 Hz, the scope's triggered time base is set at about 2 msec/div. (which shows each cycle of information consumes 16.68 msec, including 1.4 msec retrace) and amounts to all available intelligence in a single field. Two fields, of course, are required for interlace to complete one frame, of which there are 30 per second. At 15,734 Hz, the scope's time base must be adjusted for either 10 or 20 microseconds, depending on whether you wish to see just a little more than one line of horizontal sweep or several. Inverted, of course, this frequency amounts to 63.5 μsec, including 11.1 to 12 μsec for retrace (blanking) in each cycle or line. Obviously, both voltage representations are quite different since they are seen at totally unrelated frequencies, but when first viewed on poorer oscilloscopes, they can look pretty much the same.

With the development of better instruments, however, all theoretical video and sync information postulated previously by government and private industry becomes distinctly viewable and thoroughly useful. Even the composite VITS (vertical interval test signal) can be extracted with dual time base instruments and used handily for a number of subsystem evaluations. Let's take these several voltage displays one by one and see what they contain and what can be done with their specific information.

Actual Waveforms / 133

1. Vertical rate waveform photographed at 0.5V/div. vertically and 2 msec/div. horizontally (Fig. 6-25). What do we see? Just over one field of vertical information, including sync tips in negative polarity, two horizontal light lines in between indicating sync amplitude as well as black level, and then video information from black level to what appears to be a single

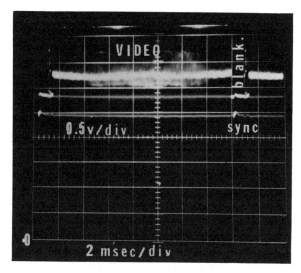

Fig. 6-24. Video at 0.5V/div. and 2 msec/div. shows just over one field of vertical information.

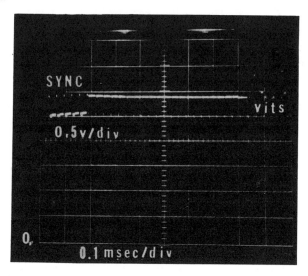

Fig. 6-25. Sync pulses and VITS at 0.1 msec/div. Zero reference shown on lowest horizontal baseline.

vertical line in each field at the end of the blanking interval denoting 0 to 100% video modulation. But, as you will discern when these two *vertical* lines are expanded horizontally by time base advance, they become the composite vertical interval test signal (VITS), which may be seen partially by exceptional oscilloscopes having single time bases and viewed completely only by good scopes with dual time bases and solid 10- or 12-kV accelerating voltages. At 0.5V/div., the signal is riding around 3V dc, and 1.25V in amplitude. Were you to switch from dc to ac amplifiers, the waveform would *fall* just over six divisions, and the actual dc reading would amount to 3.2 volts—a pretty good estimate. So you now know how to read simultaneous ac and dc voltages, speeding up service time by at least 50%.

    2. *Intermediate time base* setting of 0.1 msec (100 μsec) reveals, from left to right (Fig. 6-25) the six vertical sync pulses (positive polarity), followed by equalizing and horizontal pulses and the promised VITS display at the upper right of the picture. Naturally, if all vertical, horizontal, and equalizing pulses are correct, the receiver should have good vertical and horizontal sync. If problems do arise, then there are faults in the receiver itself *following* the video detector, but at least the intermediate frequency amplifiers (IFs) are evidently clean. We'll find out more about this shortly.

    3. *Horizontal rate displays* tell very different tales from vertical and intermediate frequencies. The time base setting here amounts to 10 μsec/div., and a single line of horizontal information and blanking amounts to 63.5 μsec. The vertical setting remains precisely the same, but you do not see the VITS information that was included in the previous display. That, remember, occupies a portion of the vertical blanking interval of a single field—lines 17 and 19, to be exact.

    In the photo (Fig. 6-26) you see a great deal of video, a horizontal sync pulse, followed by 3.58 MHz burst on the blanking interval's back porch. Note that horizontal blanking occupies just over 10 microseconds, and the sync pulse approximately half of this. Always be sure sync has correct amplitude, good rise and fall times, and proper time duration whenever checking horizontal oscillator problems. Sync pulses must occupy over 30 percent of the total waveform, whether vertical or horizontal, as they are produced via the video detector and before sync separation. Otherwise, you do, indeed, have a decided sync problem.

    4. *VITS-VIRS*, the vertical interval test and reference signals (Fig. 6-27), tell all about IF amplifiers and even luminance amplifiers, since multiburst (on the left) extends from 0.5 MHz to 4.1 MHz in six oscillatory steps, followed by a horizontal sync pulse, burst, a modified staircase, sine^2 pulses of 2T and 12.5T, a window, another horizontal sync pulse, burst, and by VIRS (the vertical information color reference signal), where color information in the broad band at the top must be in phase with burst for proper tint, and the band's amplitude should extend to the

+50 IRE point above black level and the beginning of another sync pulse. At 20 usec/div. on the B time base, you're looking at just over nine divisions, which at 63.5 usec/line amounts to some three lines of information: VITS On lines 17 and 18, and VIRS on line 19—all approved by the

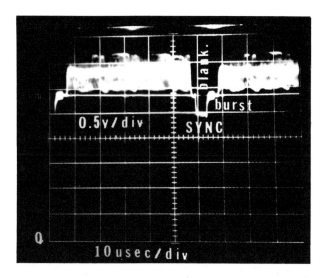

Fig. 6-26. At a faster rate of 10 usec/div., you see almost two lines of horizontal trace, along with horizontal blanking and color burst.

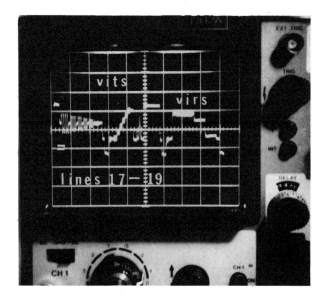

Fig. 6-27. A composite photo of lines 17, 18, and 19 as seen by any excellent oscilloscope from a broadcast signal off the air.

Federal Communications Commission and required (at least VITS is) of any TV station whose transmitter is remote from its broadcast studio.

The IFs of this receiver slope gradually from 0.5 MHz down to 3.5 MHz, show loss of some low frequency information because of horizontal sync pulse tip slant, indicate a little differential phase error and gray scale tracking in the modulated staircase, permit full passage of low-frequency luminance information in the 2T pulse, but exhibit some lack of chroma because of lower amplitude, unfilled chroma portion in the 12.5T high-frequency pulse. It's clean, with no ringing about the window, but does not deliver quite enough chroma to reach the +50 IRE Level in the VIRS portion.

Don't you agree this characterizes the IF and luminance amplifiers rather completely? In addition, if the five cycles of multiburst were not symmetrical about their centers, the receiver would most certainly be out of alignment. Naturally, we're not taking into account any antenna or transmission line effects, but that shouldn't be a problem with any reasonable exterior installation. In addition, more than one station certainly carries VITS and VIRS patterns in almost any metropolitan locale. If in doubt, try channel-to-channel comparisons.

Interestingly enough, the multiburst portion and staircase can be traced directly through the luminance amplifiers back to the cathode ray tube, where you will find most receivers delivering only 3 MHz of fine detail, but with those sets using comb filters showing between 3.5 and 4 MHz—about a 25% differential in horizontal resolution. Synchronous detection of the video amplifier also plays a considerable part in providing high resolution pictures too.

5. *Sync pulses, both horizontal and vertical*, are the final TV waveforms to be displayed in this section. The horizontal sync pulse is a single 5-μsec strobe of energy transmitted during the 11-12 μsec receiver blanking interval at the end of each horizontal line and is easily recognized; but interpreting vertical pulses is a different matter altogether, and once again you must have excellent equipment for this investigation. First, you will see a dual-trace photo (Fig. 6-28) of sync separator input (top trace) and sync separator output (bottom trace). Top trace is measured at 2V/div., while the bottom trace amounts to 20V/div.—the difference between base input and collector output of the sync separator transistor.

You will note a marked closing of the vertical blanking interval in the lower trace. With all video removed, it permits only the sync trace tip to show near the zero reference level. This means the sync separator is so biased it will conduct only on the vertical sync tips and at no other time, excluding all video and leaving only vertical sync.

Now, let's expand the time base setting just a bit to 0.1 msec (as before), and examine the lower display closely (Fig. 6-29). On the left are six good vertical sync pulses, followed by six mediocre equalizing pulses,

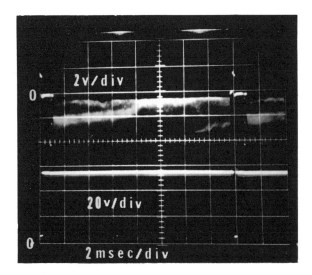

Fig. 6-28. Sync separation from any slow analog amplifier is a tricky job.

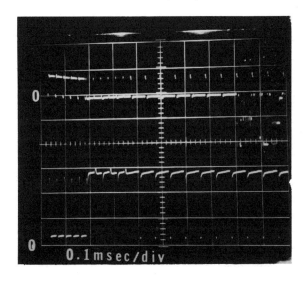

Fig. 6-29. At 0.1 msec-div. the vertical sync and equalizing pulses are quite evident.

with the remainder of the waveform's horizontal pulses normal. Observe that at the time base setting the VITS above is no longer seen in the sync separator output since it's almost a whole millisecond (10 × 0.1 msec) down the road from the vertical sync and is eliminated. Both vertical settings from the previous waveform remain at 2V and 20 V/div., respectively.

138 / Waveform Analysis

The horizontal sync pulse (Fig. 6-30), no stranger to most of you, is quite evident at an amplitude of 52 volts (bottom trace), with relatively fast rise and fall times and a 6 dB down measurement of about 5 microseconds, which is precisely what we want. The upper trace going into the sync separator has a varying amplitude of some 6 volts, depending on the quality of video present. The horizontal pulse is just visible in the initial portion of the horizontal blanking interval and is followed by a burst, then the reappearance of video. Even though this isn't the cleanest trace ever entering the sync separator, it all comes out satisfactorily in the end simply because the sync separator is so biased it conducts *only* on positive horizontal and vertical pulse tips delivered from the previous video amplifer. Of course, the time base has now been advanced to 10 μsec/div. to correspond with the 15,734 Hz horizontal frequency repetition rate. Both vertical amplifiers remain at 2 and 20 V/div as before.

Obviously there are many more common video, sync, and chroma displays in any television receiver; but for now it is important that you digest what has already been discussed, since problem displays are a working part of this discussion and will be investigated next.

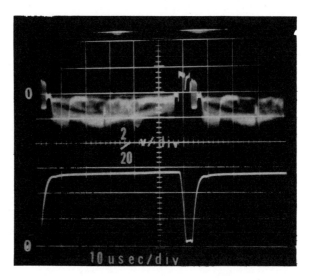

Fig. 6-30. A 10 usec/div. horizontal rate shows only single sync pulses during the blanking interval.

EXAMPLES OF DEFECTIVE WAVEFORMS AND WHY

Power Supply Ripple

Although the top Y1 trace (Fig. 6-31) is not defective, many seeing power supply input capacitative filtering for the first time actually think something is wrong. Of course, what you're actually viewing is the charge

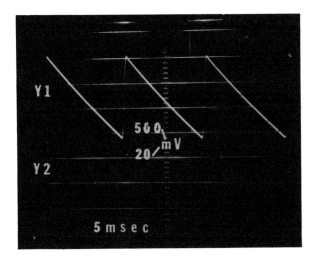

Fig. 6-31. Input capacitative filtering in power supplies produces a sawtooth, while high frequency information modulates a 60 Hz output.

and discharge of the first capacitor following half-wave rectification, which repeats at about 17-millisecond intervals. A clean trace indicates excellent initial filtering and good LC or RC time constants. This particular supply uses an inexpensive resistor in place of a better ac ripple reactance, an inductive choke.

Now note the lower Y2 trace. This is the real subject of the waveform photo. Even though somewhat indistinct, 60-Hz ripple is very real and obviously apparent. Notice, however, that it is a regular-interval smear, rather than a smear of indeterminate period irregularity. In such an instance, a higher frequency is modulating the 60Hz, rather than the reverse. So the extraneous voltage is a somewhat higher intelligence of constant frequency that's superimposed on the fully-filtered power supply.

Could you guess what it is? In this case, a multivibrator is oscillating at 80 kHz, and some of its peak amplitude transients are getting into the power supply. Can anything be done about this problem? Certainly, by the additional active or passive filter in shunt with the supply, or a decoupling network in the collectors of the relaxation oscillator. At a total of 40 mV, however, this wouldn't have enormous effect on the remainder of the system.

A little ripple here and there, nonetheless, can do a great deal by disturbing certain critical analog systems and can literally destroy the rise and fall times of sensitive digital devices. Consequently, good filtering is not only desirable in all types of circuits and systems, but absolutely necessary in most. This is why you'll find aircraft power systems at 400 Hz and many new television and industrial switch-mode power subsystems operating from 10 to 15 kHz—such frequencies are much more

140 / Waveform Analysis

effectively filtered. This is also a prime reason why Japanese consumer products have adopted active transistor-capacitor filters, so they could continue to use less expensive half-wave rectifiers and produce passable power supplies operating at 60 Hz.

Using similar vertical and horizontal oscilloscope control settings as in the previous waveform, note how the ripple at 120 Hz (fig. 6-32) reduced from some 1500 mV to 700 mV (half) in the upper waveform, and with disconnection of the additional filter, the lower waveform with more amplitude is now showing two bands of power supply interference, suggesting some kind of switch during turn-on and turn-off periods. Actually, there is no ripple, just basic clocking interference squared off top and bottom as a semiconductor relaxation oscillator swings between saturation and cutoff. Once more, high-frequency interference in the 120 Hz power supply is just a smear because of the difference in repetition rates.

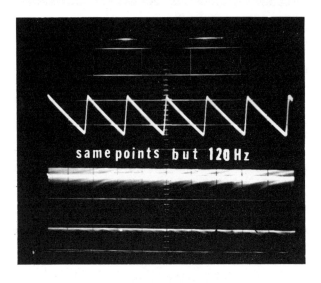

Fig. 6-32. At 120 Hz, sawtooth is reduced for better ripple filtering while power supply interference in lower trace becomes more apparent.

A Multivibrator

A multivibrator clock (totally impractical if this were a fast pulse train), illustrates the collectors of this multivibrator oscillating merrily along at a repetition rate of some 76.9 kHz (Fig. 6-33), since the time base of our oscilloscope is set at 5 usec/div. and there are some 13 microseconds per cycle. Of course, the amplitude has also changed since we no longer operate in millivolts but in solid, hefty 5V/div. Now, with baseline for each trace set on the bottom and center horizontal graticule lines, respectively, the multivibrator does, indeed, swing from zero to almost the

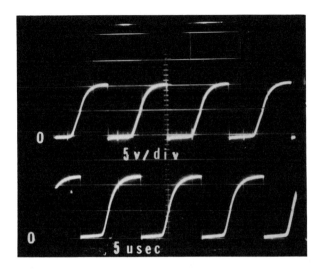

Fig. 6-33. Collector voltages illustrated by a pair of multivibrators. Voltage swing is from near 0 to 10 volts; waveform is nonsymmetrical.

power supply saturation point. Note that the bottom trace is not only inverted but differs slightly in amplitude from that at the top. But for an RC multivibrator, this one perks along pretty well. Such flip-flops, however, are both power supply and temperature sensitive, and their rise and fall trace times are obviously not symmetrical. A true set-reset flip-flop with dual triggering and no capacitors is considerably more preferable where circuit design permits.

Time Constants

The capacitor resistor RC passive device portions of any multivibrator can readily be changed so that repetition rates are easily disrupted. In Fig. 6-34, a simple 56K resistor has been placed between the collector and base of one of the oscillator transistors. *Both* sections of the multivibrator are now exhibiting actual pulses that almost look like one-shots rather than symmetrical opposites of one combined oscillatory circuit. Therefore, the rule in these dual stages is: whatever affects one side of the multivibrator affects the other, since turn-off and turn-on times must be equivalent. A small capacitance variation (Fig. 6-35) can well affect certain waveform shapings, but the frequencies, believe it or not, are just the same. What seems like large variations from one flip-flop side to the other are optical illusions, considering that one trace is exactly inverted from the other. Each cycle, however, recurs about every 15 usec, so the oscillator repetition rate is now some 66.7 kHz instead of 76.9 kHz, as it was originally, proving that RC time constants do most certainly control the frequencies (or repetition rates) of relaxation oscillators.

142 / Waveform Analysis

Fig. 6-34. Simple resistive time constant changes stretch repetition rates considerably.

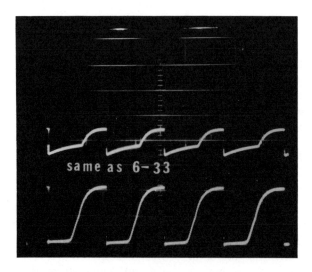

Fig. 6-35. Nonlinear sawtooth in top trace badly needs reshaping.

Nonlinearities, unfortunately, are the biggest problem we have throughout the science of waveshaping since they usually do much more harm than good. When you're attempting to derive maximum usefulness from linear representations, be very careful that capacitor couplings are valued large enough to act as short circuits (a piece of wire) and that inductors are rated in equivalent values that do not substantially interfere. A dc block by any value capacitor or an ac choke between RF or baseband and the power supply often have side effects that are not always desirable. Capacitors, of course, basically block dc and store charges, while inductors are designed for waveform peaking, ac chokes, and many varieties of auto, coupling, and isolation transformers. Sharp transients (spikes of undesirable voltage) or rectangular voltages will cause such coils to ring and produce even more aberrations that are especially annoying in video displays.

Noting that much of the previous ripple has now been substantially removed, the multivibrator and its succeeding stages may now reshape this waveform into a more symmetrical and usable form for additional circuits it may supply. For instance, by shunting a peaking inductor with a suitable capacitor, an LC time constant is formed, and a parallel turned circuit further shapes the upper voltage in Fig. 6-36 so that much more attractive representations appear in both traces. Now, with sharp rise and fall times and relatively symmetrical on and off conditions, the multivibrator output and its power supply can earn their keep in the world of active electronics. Especially observe the absence of over- and undershoots from the final rectangular voltage. This is a very clean output waveform with only a little high-frequency rounding as it reaches the low-frequency (off voltage) plateau. This means that the output switch swings fairly symmetrically between saturation and cutoff, producing highly desirable results.

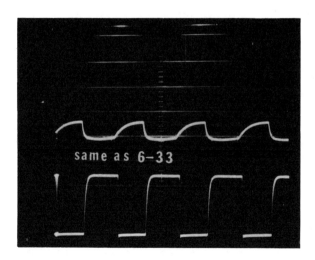

Fig. 6-36. LC parallel-tuned circuit reshapes previously distorted trace.

Video Alignments

In Fig. 6-37 you'll see the unretouched photo of a D18 Sylvania TV IF representation illustrated on the bottom trace, and the waveform of what it should really look like (or nearly so) in the top trace. The signal

144 / Waveform Analysis

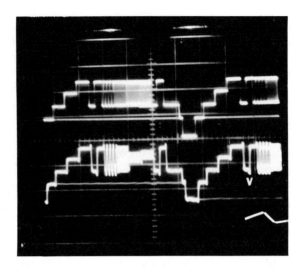

Fig. 6-37. Multiburst in TV sets or any tuned amplifier quickly defines video resolution bandpass.

generator, a Sencore VA48 that connects to any television receiver's RF terminals, is actually delivering a multiburst series of swept oscillations that begin with 0.75 MHz and extend to 3.56 MHz in four distinct series. If you think about it, these four (and they can be extended by modification to 5, at 4 MHz) actually spell out the bandpass of the tuned circuits through which they pass. Consequently, bandwidth and bandpass skirt-shaping inductors may be tuned for best response dynamically, since sync signals are already present and multiburst acts as modulated video and requires no AGC substitution for tuner or IF external control.

Of course, the bottom trace does not have nearly the full bandwidth of the upper trace that's taken directly from the VA48, and some alignment is immediately indicated because of the obviously narrow bandpass and also the position of the 3.56 MHz final multiburst that's situated between 15 and 25 percent up on the adjacent 3.02 MHz portion. The staircase that's used to check gray scale, however, is fairly linear, so a little tuning should help.

Sure enough, adjustment of the initial bandpass and one inter-coupling coil did the trick, combined with a slight readjustment of the 41.25 MHz sound trap. The results are as you see in Fig. 6-38. All multibursts have approximately the same amplitude—with only a little slope—and there are no exaggerated nonlinearities or ringing apparent. Preshoots in the staircase will not show in the picture.

Setting traps, however, is a somewhat tricky procedure and is done with preset carrier and AM modulation. Tuning then finds the point of least modulation effect. This indicates that a particular video or sound

Examples of Defective Waveforms and Why / 145

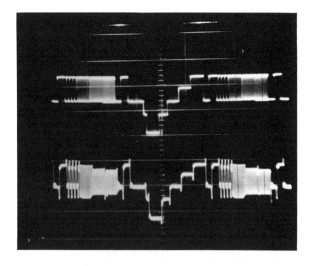

Fig. 6-38. Obviously, the multiburst-affected circuits in 1-29 required alignment, and here's the result (lower trace).

trap is set for maximum resonance. The entire "tune up" took approximately five minutes and required no more than removing the TV set's back, inserting a "cheater" cord, and looking at an oscilloscope as the three coils were turned. Where schematics aren't handy, set traps first, and then tune the bandpass transformers for maximum response.

Is this method better than using a standard sweep generator? Yes, it's much easier and very effective; but no, you aren't always completely positive of all trap settings. Each instrument, of course, has its advantages and disadvantages, just like any two similar methods in everyday life. Most people should be comfortable with either one after adequate instruction. Schooled technicians would initially prefer the sweep generator with its precise markers and externally fixed AGC, but they'd soon become used to this method for quick touchups and like it. In alignment equipment for consumer TV receivers, Sencore excels.

Sync Problems

Such faults are some of the most difficult troubles to solve, whether they occur in synchronous or asynchronous triggering among digital logic or simply in the garden variety of television receiver. Any trigger, as you must know, is usually defined as some sort of pulse that's generated from a stable, controlled source designed to coordinate timing or sweep throughout some subsystem or system. Trigger widths, amplitudes, rise and fall times, and repetition rates are all critical parameters in any sync chain that must be satisfied, especially when their rep. rates are in microseconds and nanoseconds. Video sync, fortunately,

146 / Waveform Analysis

is no faster than milliseconds and microseconds; consequently, any 15 MHz oscilloscope can handle them. Even the nine cycles of color burst on the back porch of the horizontal sync interval is well within 15 MHz range, although when its R-Y and B-Y phase references are split for chroma demodulator timing, you should have a scope time base that readily syncs and displays 200 nanoseconds/division.

Let's remember a little refresher—one from TV, that is—viewing composite video and sync out of the video detector (Y1) and the sync separator (Y2) at 1V/div. and 20V/div., respectively (Fig. 6-39). The little

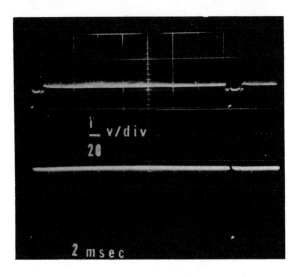

Fig. 6-39. Composite wideband sync out of the video detector.

white dots in the first and ninth graticule rows identify vertical sync tips, since our time base is set here for 2 msec/div. Each cycle, then, consumes 16.68 msec and repeats at a 59.94 Hz rep. rate. While the video detector can produce only about 2.2 volts of output at full saturation (note the dim 100 percent modulation level upper trace), the sync separator exhibits some 52 volts of high swing signal that's sufficient to drive either vacuum tubes or transistors. This, of course, is a "good" photo of what to expect when all systems are operating as they should.

Now let's begin to put some troubles into both video and sync to see what actually occurs. In the upper composite trace (Fig. 6-41), video now drops into the formerly forbidden black level and sync band, the sync pulse falls to half its former amplitude, and any picture on the tube is going to be miserable. At the same time, more video appears in the sync output—which is highly verboten—and a small, but significant transient shows just to the right of vertical sync and probably will become a disturbing sync element since it isn't supposed to be there.

Examples of Defective Waveforms and Why / 147

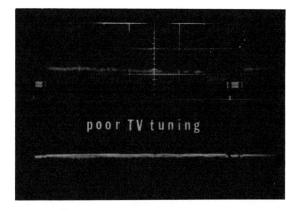

Fig. 6-40 Sync tips on the lower trace are somewhat elusive, but they're there nonetheless.

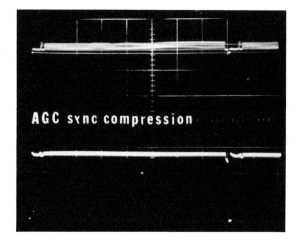

Fig. 6-41. Video weakens and compresses in top trace, while video begins to show in the lower sync display.

In the next problem, Fig. 6-41, the AGC forward biases the IF transistors very nearly into saturation, composite video becomes compressed, and the sync notch deepens since the sync pulse itself no longer occupies 30 percent of the composite video waveform. Would you expect adequate sync to pass through to the vertical oscillator under such conditions? Of course not. Such sync would become very unstable, the picture could roll or shake, and overall viewing conditions would become quite unacceptable. Whether TV or not, any sync pulse that's clipped or distorted can't possibly do its intended job if either symmetry or amplitude is reduced below working levels. So whenever you're looking for poor overall system

148 / Waveform Analysis

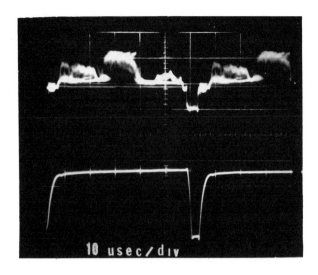

Fig. 6-42. A great deal of sync compression shows up in bottom trace as sync distortion along the 8th vertical graticule line.

timing or outright sync instability, go back to sync origin and see whether all generating waveforms are adequate. If not, first be sure some feedback difficulty isn't your auxiliary problem, then proceed to the main sync pulse or timing oscillator and repair whatever may be the trouble. Should interference be filtering back, find that malfunction first and then proceed with the sync source, if necessary.

Horizontal sync is not as critical as vertical sync because of the automatic frequency control to which each horizontal oscillator is subject, and the more rapid 15,734 Hz repetition rate makes control considerably easier. And even though the 1V/div. and 20V/div. out of the video detector and sync separator remain, the initial horizontal sync pulse must remain squared off (as shown in the top trace of Fig. 6-42), and the bottom trace must deliver this sync to the AFC transistors or diodes for adequate flyback and sync pulse comparison so the horizontal oscillator is always maintained in sync. The leading edge of this sync pulse *must* stay sharp, and its 6 dB down pulse width cannot exceed more than 5 usec. Note good amplitude burst on the back porch of the sync pulse during the transmitted 11.11 usec horizontal blanking interval, which is often stretched by ordinary color receivers to 12 usec, cutting off just a bit of horizontal scan. As usual, this horizontal sync pulse is shaped by an RC integrator and the capacitor may neither change value nor develop leakage if the waveform is to maintain full integrity.

Examples of Defective Waveforms and Why / 149

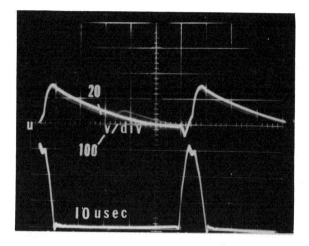

Fig. 6-43. Sync pulse edges must remain sharp and of the proper amplitude in any system that uses them for turn on.

AFC Phase Control

Control of the horizontal oscillator is one of the trickier circuits in video because without it horizontal sync could not be continuously maintained. Figure 6-43 illustrates the incoming sync pulse at 20 volts/division and the horizontal feedback pulse (not yet integrated) at 100 volts/division. Observe that the conduction time (low) of the sync pulse corresponds to the

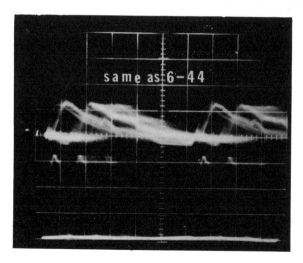

Fig. 6-44. Coincident sync, especially in phase comparators, must be precise.

leading edge of the horizontal pulse (output off time); and between the two, they deliver any needed correction in the form of DC voltage to the horizontal oscillator. Is it any wonder that when external conditions or an internal breakdown triggers this incoming sync pulse erratically that the results shown in Fig. 6-44 occur? You'd have a hard time maintaining horizontal sync with this one since AFC, oscillator, horizontal output, damper, and even high voltage (especially regulation) are all confused by such jittery behavior.

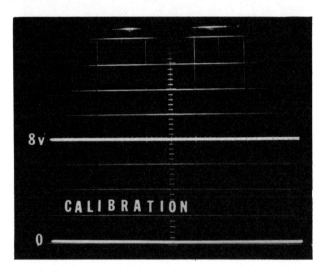

Fig. 6-45. Use a DC source and digital voltmeter for routine vertical amplifier calibration.

You should be well aware by now that sync difficulties may mirror other problems, can often generate their own, and will cause all sorts of havoc when either phase or frequencies are incorrect. Whenever there are raster jitter or apparent sync problems in any video display, immediately look to the video output and sync sections for keys to your difficulties. In these instances, signal tracing is the only solution, and the better the oscilloscope, the quicker repairs can be made.

SCOPE CALIBRATION

You might have lots of extra money for time mark and constant voltage generators—and then, again, you may not. Perhaps knowing a couple of first class substitutes that are almost as effective might be worthwhile.

Vertical Calibration

Unless your vertical attenuators are way off the mark, a constant voltage generator isn't really necessary. You can readily substitute a tenuator setting of 2 volts/div. (Fig. 6-45) the center of the graticule, of mediocre power source and digital voltmeter instead. At a vertical at- course, would then be 8 volts, etc. In this way, each attenuator setting can be checked by voltmeter for overall calibration. If you wish to see if such vertical amplifiers are also linear with respect to phase, then introduce a square wave such as shown in the top trace of Fig. 6-46. A bad test probe or capacitor among the vertical attenuators will show up immediately, the same as a partially disconnected common (ground). But do vary the frequency of your (linear, we hope) generator so that these vertical amplifiers have a comprehensive test.

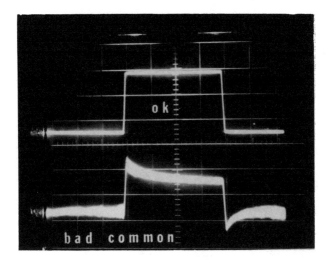

Fig. 6-46. Checking vertical amplifiers and probes with a square wave.

Finally, don't forget your low capacitance probes. They, too, must be squared off with rectangular waves for non-distorted, full frequency response.

Horizontal Calibration

Here we're really speaking of time base since there is no horizontal amplifier calibration in any but the oldest oscilloscopes, most of which are now extinct. The easiest method (Fig. 6-47) would be the use of a signal generator and period (inverse of frequency) of an accurate electronic counter. Start by feeding a 1 kHz (10^3) signal into the scope's vertical amplifer(s) and adjust its time base for a millisecond (10^{-3}). If at least eight of the resulting pulses are on the vertical marks, that time base is correctly calibrated. Try other frequencies at 10^2 (100 Hz) and 1 and 10 MHz, respectively (10^6 and 10×10^6). If there are no obvious errors, the oscilloscope is adequately calibrated. If not, begin again with the 1 msec

152 / Waveform Analysis

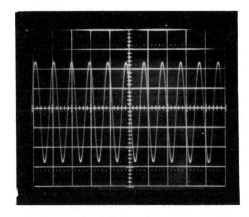

Fig. 6-47. Why not calibrate time base periods too? Use a counter and sine wave signal generator—that's all!

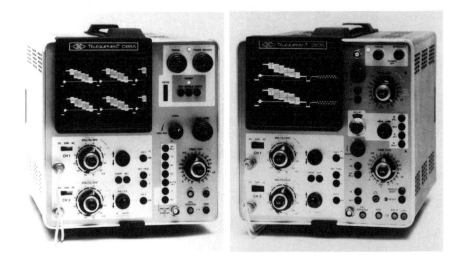

Fig. 6-48. D66A and D67A excellent Telequipment oscilloscopes, with vertical deflection factors of from 1 mV to 50 V/div. and time bases from 2 seconds to 100 and 200 nanoseconds, respectively, that displayed all of the foregoing waveforms. Bandpass is 3 dB down at 27 MHz (actually measured).

range and work up, adjusting for best overall calibration. Where one time base setting cannot be calibrated within specifications, you have an obvious problem and the instrument requires repairs. The same conditions, obviously, hold true when no single time base range can be successfully calibrated.

Scope Calibration / 153

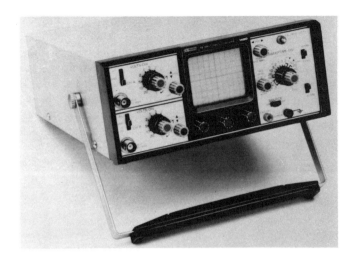

Fig. 6-49. B & K Precision Model 1420 15 MHz portable miniscope. Battery optional.

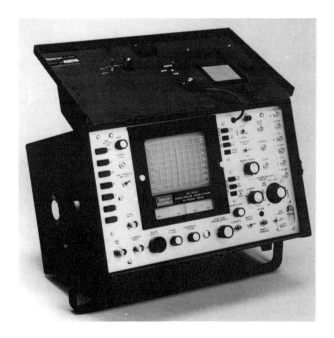

Fig. 6-50. New VSM-2A and VSM-5B battery or AC-operated Texscan portable spectrum analyzers with phase lock stability, variable sweep, harmonic markers, and frequency ranges to 1 GHz and 450 MHz, respectively. (Courtesy of Tekscan Corp., Indianapolis, Ind.)

Usually, service-type oscilloscopes have only two calibration potentiometers for each overall calibration, the intermediate ranges being set at the factory. Therefore, your service manual should show these variable accuracy controls and the process will turn out to be quite easy, especially after your second and third attempts. In this intance, use the 1st and 9th vertical lines for your calibration market points and note there is a deviation of some 0.3 higher in frequency than perfection. Consequently, error amounts to: 8 divisions divided into 0.3 (8/0.3) which amounts to 0.375 × 100 = 3.75%. If that isn't good enough, find the calibrate potentiometer and go to work. But do look at all other frequencies (or periods) before you consider the job well done.

INDEX

AC-DC measurements, 13
AFC phase control, 149-50
AM distortion, 41
AM modulation, 40-41
Amplitude measurements, 36-37
Asynchronous sampling, 60-61
Automatic trace erasure, 72
Auto Mode, 9
Azimuth Recording, 103

B & K-Precision, 89, 92
Bar Sweep mode, 49
Baseband, 102
Bessel Null Chart, 44-46
BETA systems, 101
   drawbacks of, 104-106
Binary table display, 62-63
Biomation (Gould), 53
Bistable storage, 73
Bit pattern triggering, 61
Borchert, Marshall 65
Bright bistable, 73
B-Y patterns, 99

Calibration, 13-14
Cassettes (video), 100-11
Cathode ray tube displays, 62-64
Chroma, 96, 97
   vectorscope, 90-94
   see also Color; Luminance

156 / Index

Chroma Sweep mode, 49, 51
Clock timing, 106–107
Closed circuit TV systems, 91
Color
    cassettes, 109
    gaited rainbow color bar generators, 96–99
    vector pattern, 94–96
    vectorscopes, 23–26
    see also Chroma; Luminance
Color bar generators, 96–99
Composite video waveforms, 132–38
Computers, 4
Cursor, 57

DC offsets, 16–17
Deviation ratio, 44
Digital analysis, 58–59
Digital logic, 4
Digital multimeters, 4
Digitizing, 53
Diode demodulator probes, 11
Dual beam oscilloscopes, 7
Dual trace oscilloscopes, 7

Electronics Industries Association (EIA), 121
Envelope diode detector, 11
External sync, 9

Fast mesh transfer, 75
Fast multimode storage, 75–76
Federal Communications Commission, 79
FET probes, 11
Field Effect Transistor probe, 10
FM modulation, 44–45
FM spectrum measurements, 44–46
Fourier spectrum, 40
4-trace oscilloscopes, 7
Frequency bases, 38–40
Frequency domain reflectometry, 87
Frequency modulation, 44
Frequency response difference, 3
Fundamental waveshapes, 125

Gaited rainbow color bar generators, 96-99
Glitches, 76, 127
Gould, Inc. 72

Hardware systems, 56
Harmonic distortion, 47, 48
Harmonics, 42-43
Hewlett-Packard, 35
   digital analysis, 58-59
   logic analyzers, 53, 54-55, 56, 63
   storage oscilloscopes, 72, 74
Horizontal amplifiers, 3
Horizontal calibration, 151-53
Horizontal rate displays, 134

Institute of Radio Engineers (IRE), 95
Intermediate time base, 134
Intermodulation distortion, 47-48
IRE (Institute of Radio Engineers), 95

Jitter, 33
JVC, 101

Lissajous oscilloscope patterns, 20, 130
Logic analyzers, 53-69
   defined, 54-56
   operation of, 60-65
   testing procedures, 56-59
   troubleshooting, 65-69
Luminance, 96, 97, 109
   see also Chroma; Color

Magnavox, 101
Magnetic wraparound, 10
Map display, 63-65
Matrix algebra, 21
Matshushita, 101
Microprocessors, 4
   logic analyzers, 53
   troubleshooting, 65
Modulation index, 44

Multivibrator, 140-41

National Television System Committee (NTSC) 90-92, 94, 96, 102, 108
"Nested loops," 56
Null points, 46

Oscillator distortion, 43
Oscillograph, 3
Oscilloscopes
  AC-DC measurements, 13
  applications, 15-20
  basics of, 3-6
  calibration, 13-14
  defined, 3
  probes, 10-12
  professional, 28-33
  spectrum analyzer contrasted, 34
  time base, 9-10
  vectorscopes, 21-28
  vertical amplifiers, 6-9
Oscilloscopes (sampling), 82-83
  storage scopes and, 70
Oscilloscopes (storage), 71-82
  fast scopes, 80-82
  features of, 73-76
  function of, 76-80
  reasons for, 73
  sampling scopes and, 70

Planar networks, 56
Probes, 10-12
Power output test, 47
Power supply ripple, 138-40
Professional oscilloscopes, 28-33

Quasar, 101

Ramp waveforms, 131
Random access memories (RAM), 60, 106-107
Rectangular voltages, 125-28
Rectangular waves, 113-14

Resistive isolation direct probes, 10-11
Resolution 37-38
R-Y patterns, 98-99

Sampling oscilloscopes, 82-83
Sanyo, 101
Sawtooth waveforms, 131
   analysis, 115-16
Semiconductors, 4
Sencore, 33, 49, 50, 89, 96, 144, 145
"Sidelock" color bar generators, 96-99
Signal-to-noise (S/N) ratio, 43, 48-52
Sine wave, 15-20
Sinusoidal voltages, 128-31
Sinusoidal waveshapes, 13
S/N ratio, *see* Signal-to-noise (S/N) ratio
Software systems, 56
Sony, 101
Spectrum analyzers, 34-52
   AM spectrum measurements, 40-43
   applications of, 36-40
   FM spectrum measurements, 44-46
   omni-checks, 47-52
   oscilloscopes contrasted, 34
   varieties of, 34-36
Split beam oscilloscopes, 7
Spurs, 77
Square waves, 17-18
   analysis, 113-14
Staircases, 116
State analyzers, 55-56
Storage oscilloscopes, *see* Oscilloscopes (storage)
Sylvania, 143
Synchronous sampling, 60, 61
Sync problems, 145-49
Sync pulses, 136-37

TDR, *see* Time-domain reflectometry
Tektronix, 13, 29, 30, 31, 35, 36, 44, 46, 47, 48, 49
   logic analyzers, 53-54, 55, 56, 57, 59, 60, 61, 65-69
   sampling oscilloscopes, 82
   storage oscilloscopes, 71, 72, 74, 75, 77, 80, 81
   time domain reflectometry, 84

Tektronix *"Contd."*
  vectorscope, 90, 91
  VIRS-VITS generators, 122
Telequipment, 92
Teletext, 79, 122
10:1 low capacitance probe, 10-11
Time base, 9-10
  spectrum analyzers, 38-40
Time constants, 141-43
Time domain reflectometry, 70, 83-88
Timing analyzers, 55
Timing display, 62
Triangular waves, 114-15
Triggering, 61
Troubleshooting
  logic analyzers, 65-69
  microprocessors, 55
  vectorscopes, 26-28
True RMS (root-mean-square) reading meters, 4
Tuners, 102

U.S. air cable transmission system, 79

Variable persistence storage, 74-75
Vector patterns, 94-96
Vectorscopes, 3, 4, 21-28
  chroma, 90-94
  color TV, 23-26
  troubleshooting, 26-28
  verifying results, 21-22
Vernier (fine) adjustments, 9-10
Vertical amplifiers, 3, 6-9, 13
Vertical calibration, 150-51
Vertical rate waveform, 133-34
Video alignments, 143-45
Video cassettes, 100-11
Video Home System (VHS), 101, 104-106
Video terminals, 89-102
Viewdata, 79
VIRS (vertical color reference signal), 120-22, 134-36
VITS (vertical interval test signal), 117-20, 132, 134-36, 137
Voltage comparator, 60
Voltage-controlled oscillator (VCO), 104

Voltage levels, 92
Vuedata, 122

Waveform analysis 112-53
    actual waveforms, 122-38
    basic configurations, 112-16
    defective examples, 138-50
    scope calibration, 150-53
    VIRS, 120-22
    VITS, 117-20
Word recognition, 61

X-axis (intensity) modulation, 8
X-Y oscilloscope, 91, 94-95
X-Y pattern, 123

Zenith, 101
Zenith-Sony, 103, 106